Essential Semiconductor Laser Device Physics (Second Edition)

Online at: https://doi.org/10.1088/978-0-7503-6417-1

Essential Semiconductor Laser Device Physics (Second Edition)

A F J Levi

Professor of Electrical and Computer Engineering and Physics and Astronomy,
University of Southern California, Los Angeles, CA, USA

IOP Publishing, Bristol, UK

ISBN   978-0-7503-6417-1 (ebook)
ISBN   978-0-7503-6413-3 (print)
ISBN   978-0-7503-6414-0 (myPrint)
ISBN   978-0-7503-6416-4 (mobi)

DOI   10.1088/978-0-7503-6417-1

Version: 20250501

IOP ebooks

British Library Cataloguing-in-Publication Data: A catalogue record for this book is available from the British Library.

Published by IOP Publishing, wholly owned by The Institute of Physics, London

IOP Publishing, No.2 The Distillery, Glassfields, Avon Street, Bristol, BS2 0GR, UK

US Office: IOP Publishing, Inc., 190 North Independence Mall West, Suite 601, Philadelphia, PA 19106, USA

Cover image: Transient probability of quantized carrier and photon number in a mesoscale laser diode subject to a step change in injection current. Image credit: K Roy-Choudhury, A F J Levi.

In memory of Professor Chia Wei (Wade) Hsu

1988–2024

Contents

Preface

The semiconductor laser diode is a key technology enabling the continued expansion of advanced global communications, data centers, and large-scale computing infrastructure. In addition to these commercially successful applications, researchers and technologists are actively pursuing a broad range of potentially impactful emerging capabilities, from laser-based automobile collision avoidance to secure quantum key distribution.

This book describes the key aspects of semiconductor laser device physics and principles of laser operation. It is an accessible, convenient reference and provides essential knowledge that may be easily understood before exploring more sophisticated device concepts. The contents serve as an essential foundation for scientists and engineers about how semiconductor lasers work and the fundamentals determining their behavior without requiring highly specialized and detailed study. The book provides the necessary knowledge to develop novel devices and system configurations exploiting the same fundamental physics of light–matter interaction in semiconductor lasers.

New material in the second edition includes expanded and improved descriptions of basic concepts and practical applications. A chapter on quantum effects in small semiconductor lasers has been included, and appendices provide helpful supplemental material.

A F J Levi
California, 2025

Author biography

A F J Levi

Tony Levi joined the faculty of the University of Southern California in mid-1993 after working for 10 years at AT&T Bell Laboratories, Murray Hill, NJ, USA. He invented hot-electron spectroscopy, discovered ballistic electron transport in heterostructure bipolar transistors, demonstrated room temperature operation of unipolar transistors with ballistic electron transport, created the first microdisk laser, and carried out groundbreaking work in the optimal design of small electronic and photonic systems. His current research interests include device physics at the classical–quantum boundary, system engineering and integration, high-performance electronics, and optimization in system design. To date, he has published numerous scientific papers, several book chapters, and is the author of the books *Applied Quantum Mechanics*, currently in its third edition, *Essential Classical Mechanics for Device Physics*, *Essential Semiconductor Laser Device Physics*, *Essential Electron Transport for Device Physics*, co-editor of the book *Optimal Device Design*, and holds 17 US patents.

List of symbols

$\lvert 1 \rangle$	electronic state 1
$\hbar$	Planck's constant
e	electron charge
c	speed of light
ε_0	permittivity of a vacuum
ε_{r}	relative permittivity in a medium
$\varepsilon_{\mathrm{r0}}$	low-frequency relative permittivity in a medium
$\varepsilon_{\mathrm{r\infty}}$	high-frequency relative permittivity in a medium
μ_0	permeability of a vacuum
μ_{r}	relative permeability in a medium
T	absolute temperature
k_{B}	Boltzmann constant
f_b	Fermi–Dirac distribution in band b
g_{B}	Bose occupation factor
μ	chemical potential
E_{F}	Fermi energy, $E_{\mathrm{F}} = \mu(T = 0\ \mathrm{K})$
$\Delta\mu$	difference in quasi-chemical potential
P	spectral density
m_0	bare electron mass
m^*	effective mass
m_{e}^*	effective electron mass
m_{hh}^*	effective heavy hole mass
m_b^*	effective mass in band b
m_{r}^*	reduced effective mass
$\mathbf{k}$	wave vector; in one dimension this is wave number k
k	wave number; signed scalar value of the electron wave vector
k_{F}	Fermi wave number
$\gamma_{\mathbf{k}}$	electron scattering rate
$E_{\mathbf{k}}$	electron energy
v_{g}	group velocity
v_{p}	phase velocity
E_{g}	bandgap energy
D_3	density of electron states in three dimensions
n	carrier density
n_b	carrier density in band b
n_{th}	carrier density at laser threshold
n_{ot}	carrier density at optical transparency
I_{inj}	injection current
I_{th}	threshold current
I_{fl}	contribution to the current from subthreshold photon fluctuations
V_{bias}	bias voltage
W_{21}	matrix element coupling state $\lvert 2 \rangle$ to $\lvert 1 \rangle$
B_{21}	stimulated transition coefficient between electronic state $\lvert 2 \rangle$ and $\lvert 1 \rangle$
A_{21}	spontaneous transition coefficient from electronic state $\lvert 2 \rangle$ to $\lvert 1 \rangle$
R_{12}^{stim}	stimulated emission rate from electronic state $\lvert 1 \rangle$ to $\lvert 2 \rangle$
R_{21}^{stim}	stimulated emission rate from electronic state $\lvert 2 \rangle$ to $\lvert 1 \rangle$

R_{21}^{spon}	spontaneous emission rate from electronic state $|2\rangle$ to $|1\rangle$
r_{spon}	total spontaneous emission rate
$\mathbf{E}$	electric field
$\mathbf{E}_{opt}$	classical optical electric field
α_{opt}	optical absorption
g_{opt}	optical gain
g_0	material-dependent optical gain coefficient
r_0	material-dependent optical gain coefficient
k_{ph}	photon wave number
$\mathbf{k}_{opt}$	wave vector of classical optical electric field
n_r	refractive index
ω	radial frequency
$\Delta\omega$	difference in radial frequency
ν	frequency
ν_{Re}	real part of frequency
ν_{Im}	imaginary part of frequency
ν_0	optical frequency
$\Delta\nu$	difference in frequency
$\Delta\nu_{FWHM}$	full-width-at-half-maximum of spectral linewidth
γ_{opt}	optical linewidth
α	linewidth enhancement factor
Γ_{opt}	optical confinement factor
β	spontaneous emission factor; fraction of total spontaneous emission feeding into a laser mode
L_{out}	light output
λ	wavelength
$\Delta\lambda$	difference in wavelength
$S_{Poynting}$	magnitude of the Poynting vector
S	photon density in lasing mode
S_0	average value of optical intensity
τ_{RT}	round-trip time
τ_r	rise time
τ_{bit}	bit time
τ_{ph}	photon lifetime
τ_n	carrier lifetime; inverse carrier lifetime $1/\tau_n = A_{nr} + Bn + Cn^2$.
A_{nr}	nonradiative recombination rate
B	spontaneous emission rate
C	higher-order term contributing to nonradiative recombination
κ	total photon loss rate, $1/\tau_{ph}$
α_i	internal photon loss rate, $1/\tau_{int}$
α_m	mirror photon loss rate, $1/\tau_{mirror}$
G	optical gain in lasing mode
G_{slope}	differential optical gain with respect to carrier density
ε_{bulk}	gain saturation coefficient in a bulk semiconductor
ε_{QW}	gain saturation coefficient in a semiconductor quantum well
D_3^{ph}	photon density of states in three dimensions
L	crystal lattice constant
L_C	cavity length
L_m	dielectric mirror thickness
V_{vol}	volume; a cube of side L_{vol} has volume V_{vol}

I_{opt}	optical intensity
$\mathcal{F}$	finesse
Q	quality factor
τ_Q	ring-down time constant of a resonator with quality factor Q
r_{ph}	reflection amplitude
t_{ph}	transmission amplitude
t_a	active region thickness
n_a	refractive index of the active region
n_c	refractive index of the optical confinement layer
$f(t, y)$	function of t and y
$P_{n,s}$	probability of positive semi-definite integer n excitations and positive semi-definite integer s photons
P_r	pump rate
P_{norm}	normalized pump rate
N_e	number of emitters
TE	transverse electric
TM	transverse magnetic
LD	laser diode
LED	light-emitting diode
NRZ	non-return to zero
BER	bit error ratio
SNR	signal power to noise ratio
RIN	relative intensity noise

IOP Publishing

Essential Semiconductor Laser Device Physics (Second Edition)

A F J Levi

Chapter 1

Semiconductor band structure and heterostructures

This chapter reviews some aspects of semiconductor material science that directly supports the design and manufacture of laser diodes. The Schrödinger equation for a single electron in the periodic potential of a crystal, Bloch's theorem, the Bloch wave vector, band gaps in the electron dispersion relation, and the origin of complex band structure are described. An introduction to semiconductor heterostructures, substitutional doping, thermal equilibrium, particle distribution functions, and the heterostructure diode is also provided.

Before describing how a laser diode works, it is worth recognizing the contribution of the precision growth of single-crystal semiconductor heterostructures to the design and manufacture of efficient laser diodes. It is also helpful to understand the material science that directly supports laser diode technology. Both manufacturing and material science are important enablers of a systematic approach to optimal device design—and, of course, semiconductor device physics is another essential element aiding technological development.

It turns out that, to some extent, elementary quantum mechanics can be used to explain how semiconductor band structure and heterostructures contribute to device performance [1]. The critical features of *bulk* semiconductor band structure that emerge include crystal symmetry, the energy dispersion of electron propagating states with wave character, the density of electronic states, and the existence of an energy band gap for which there are no electron propagating states. This and related device physics phenomena are described in the following.

1.1 Atom shape and crystal structure

If little was known about atoms being condensed from a gas or a liquid into a crystalline solid, it seems reasonable to assume the lowest energy state of this system is stable and has a high density (the atoms are close together). If the atoms are

doi:10.1088/978-0-7503-6417-1ch1
1-1

identical noninteracting spheres, then high density can be achieved by placing atoms in stacked hexagonal close-packed planes. In this case, the shape determines the close-packed arrangement (a steric effect).

However, it is known that bulk semiconductors such as silicon (Si) and gallium arsenide (GaAs) have different crystal structures. Si is a group IV element that forms a diamond crystal structure. An example of a III–V compound semiconductor is GaAs, which forms a zinc blende crystal structure.

The outer electron density surrounding an atom nucleus is usually not simply spherical but has angular dependence, and it is this electronic shape that contributes to determining the type of crystal lattice that forms. Si is a tetrahedrally coordinated solid with a diamond crystal structure because the four outermost electrons of the atom are available for chemical bonding, and the probability distribution for those electrons can take on a shape resulting in a specific crystal structure that minimizes the total electronic energy in the system. Bonds consisting of hybrid orbitals favor particular bonding directions. A hexagonal close-packed structure does not form because the spatial shape of the electron density around each atom does not minimize the overall energy for that configuration.

To appreciate where the shape of atoms comes from, consider the hydrogen atom which consists of a negatively charged electron and a nucleus that is a positively charged proton. The Coulomb interaction between these equal but oppositely charged particles creates a spherically symmetric potential. The natural coordinate system is spherical, as illustrated in figure 1.1. The position of the electron relative to the proton is given by radial coordinate r and angles θ and ϕ. Separating the center-of-mass and relative motion of the electron from the much heavier proton, the quantized angular momentum of the single electron is described in terms of spherical harmonics, Y_l^m, with integer quantum numbers $l = 0, 1, 2, \ldots$ and $-l \leqslant m \leqslant l$. It is this quantization of electron orbitals that gives rise to the angular dependence of electron probability density in an atom.

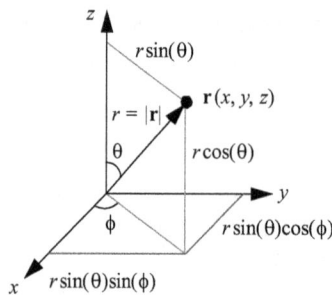

Figure 1.1. Illustration of a spherical coordinate system relative to Cartesian axes x, y, and z. Position vector $\mathbf{r}$ is described using the radial coordinate $r = |\mathbf{r}|$, the polar angle is θ, and the azimuthal angle is ϕ.

1.1.1 The spherical harmonics

The first few spherical harmonics $l = 0$ (s-orbital), $l = 1$ (p-orbitals), and $l = 2$ (d-orbitals) are

$$Y_0^0 = \left(\frac{1}{4\pi}\right)^{1/2} \tag{1.1}$$

$$Y_1^0 = \frac{1}{2}\left(\frac{3}{\pi}\right)^{1/2} \cos(\theta) \tag{1.2}$$

$$Y_1^{\pm 1} = \mp\frac{1}{2}\left(\frac{3}{2\pi}\right)^{1/2} \sin(\theta)e^{\pm i\phi} \tag{1.3}$$

$$Y_2^0 = \frac{1}{4}\left(\frac{5}{\pi}\right)^{1/2} (3\cos^2(\theta) - 1) \tag{1.4}$$

$$Y_2^{\pm 1} = \mp\frac{1}{2}\left(\frac{15}{2\pi}\right)^{1/2} \sin(\theta) \cos(\theta)e^{\pm i\phi} \tag{1.5}$$

$$Y_2^{\pm 2} = \frac{1}{4}\left(\frac{15}{2\pi}\right)^{1/2} \sin^2(\theta)e^{\pm 2i\phi}. \tag{1.6}$$

Plots of the magnitude of these spherical harmonics, $|Y_l^m|$, along with the names of the orbitals, are shown in figure 1.2.

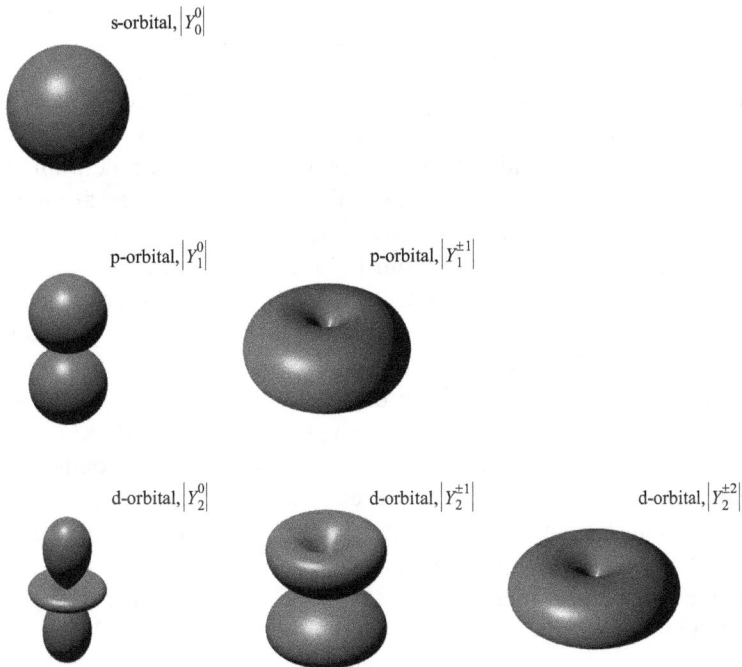

Figure 1.2. Plots of the magnitude of the first few spherical harmonics, $|Y_l^m|$, with the name of the orbital indicated. The Cartesian coordinates x, y and z are as shown in figure 1.1, with the origin at the center of each plot.

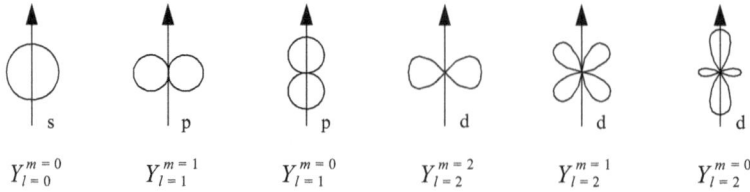

$$Y_{l=0}^{m=0} \qquad Y_{l=1}^{m=1} \qquad Y_{l=1}^{m=0} \qquad Y_{l=2}^{m=2} \qquad Y_{l=2}^{m=1} \qquad Y_{l=2}^{m=0}$$

Figure 1.3. Illustration of the first few spherical harmonics by plotting the value of the absolute value of Y_l^m as a function of the polar angle θ in a plane through the z-axis.

These figures may be related to Y_l^m plotted in figure 1.3 as a function of polar angle θ in a plane through the z-axis.

Spherical harmonics and the quantization of angular momentum play an important role in determining the eigenvalues and eigenfunctions of many atomic and molecular systems. For these systems, angular momentum contributes to the potential seen by electrons and the shape of the electron density distribution about the nucleus of any given atom.

1.2 Atomic orbitals and hybridization

The orthonormal basis of stationary electron states of the hydrogen atom is separable into radial ($R_{nl}(r)$) and angular functions ($Y_l^m(\theta, \phi)$) such that

$$\psi_{nlm}(r, \theta, \phi) = R_{nl}(r) Y_l^m(\theta, \phi). \tag{1.7}$$

These atomic orbitals have quantum numbers n, l, and m that determine the electron's quantized energy eigenvalue E_n, the square of the angular momentum eigenvalue, $\hbar^2 l(l+1)$, and the z component of angular momentum eigenvalue, $\hbar m$. The bound-state electron energy eigenvalues in the hydrogen atom are discrete and only depend on the non-zero positive integer principal quantum number, n. The orbital angular momentum quantum number is an integer $0 \leqslant l \leqslant n-1$, and the quantum number m is an integer such that $-l \leqslant m \leqslant l$. An electron bound-state in hydrogen with a principal quantum number n has energy degeneracy n^2.

It is an intrinsic feature of quantum mechanics that new stationary states can be formed from a linear superposition of degenerate $\psi_{nlm}(r, \theta, \phi)$ states. Such a linear superposition of orbitals with the *same* value of n but *different* l and m is called a *hybrid orbital*. The creation of hybrid orbitals allows electron density to be enhanced in a particular spatial direction compared to the pure atomic orbitals from which it is formed. It is in this way that directional bonds and the resulting geometry of many molecules and crystals can be explained.

As an example, consider the $n = 2$, $l = 1$ (2p) states of the hydrogen atom. They have orthogonal atomic orbitals

$$\phi_{2,1,1} = -\frac{1}{2}\sqrt{\frac{3}{2\pi}} R_{2,1}(r) \sin(\theta) \, e^{i\phi} \tag{1.8}$$

$$\phi_{2,1,0} = \frac{1}{2}\sqrt{\frac{3}{\pi}} R_{2,1}(r) \cos(\theta) \qquad (1.9)$$

$$\phi_{2,1,-1} = \frac{1}{2}\sqrt{\frac{3}{2\pi}} R_{2,1}(r) \sin(\theta)\, e^{-i\phi}. \qquad (1.10)$$

These can be rewritten as linear combinations of orbitals to form p_x, p_y, and p_z orbitals that are real functions of r, θ, ϕ. They are orthogonal and span the same part of Hilbert space:

$$\phi_{2p_x} = \frac{-1}{\sqrt{2}}(\phi_{2,1,1} - \phi_{2,1,-1}) = \sqrt{\frac{3}{4\pi}} R_{2,1}(r)\frac{x}{r} \qquad (1.11)$$

$$\phi_{2p_y} = \frac{i}{\sqrt{2}}(\phi_{2,1,1} + \phi_{2,1,-1}) = \sqrt{\frac{3}{4\pi}} R_{2,1}(r)\frac{y}{r} \qquad (1.12)$$

$$\phi_{2p_z} = \frac{1}{\sqrt{2}}\phi_{2,1,0} = \sqrt{\frac{3}{4\pi}} R_{2,1}(r)\frac{z}{r}. \qquad (1.13)$$

These functions may be used to illustrate the formation of hybrid orbitals. Consider the carbon atom with four outer shell electrons in the $2s^2 2p^2$ configuration. Hybrid orbitals for these four electrons can be formed using the p_x, p_y, and p_z orbitals and the 2s hydrogenic orbital. These hybrid orbitals are

$$\psi_1 = \frac{1}{\sqrt{4}}(\phi_{2s} + \phi_{2p_x} + \phi_{2p_y} + \phi_{2p_z}) \qquad (1.14)$$

$$\psi_2 = \frac{1}{\sqrt{4}}(\phi_{2s} + \phi_{2p_x} - \phi_{2p_y} + \phi_{2p_z}) \qquad (1.15)$$

$$\psi_3 = \frac{1}{\sqrt{4}}(\phi_{2s} + \phi_{2p_x} - \phi_{2p_y} - \phi_{2p_z}) \qquad (1.16)$$

$$\psi_4 = \frac{1}{\sqrt{4}}(\phi_{2s} - \phi_{2p_x} + \phi_{2p_y} + \phi_{2p_z}). \qquad (1.17)$$

The electron probability density has maxima along the four tetrahedral directions, which in Cartesian coordinates are $(1, 1, 1)$, $(-1, -1, 1)$, $(1, -1, -1)$, and $(-1, 1, -1)$. The methane molecule, CH_4, has a hydrogen atom covalently bonded by the electron probability maximum in each tetrahedral direction of the outer shell electrons in the carbon atom. Figure 1.4 shows different ways of visualizing the covalently bonded methane molecule.

The directionality of bonds explains why crystals have the symmetric structure they do. In three-dimensional space, the location of atoms at crystal lattice sites can be described in terms of just seven crystal systems. A fixed group of atoms called a

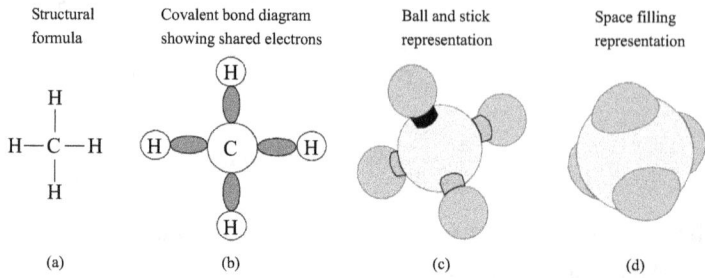

Figure 1.4. Different ways of representing the structure of a tetrahedrally coordinated methane molecule, CH_4. (a) The structural formula, (b) covalent bond diagram, (c) ball and stick representation, and (d) space-filling representation.

basis is associated with each lattice point. The crystal is exactly filled when basis atoms are placed at each lattice point. See appendix B.

1.3 The one-electron Schrödinger equation

The solution to the non-relativistic one-electron time-independent Schrödinger equation

$$\hat{H}\psi(x) = \left(\frac{-\hbar^2}{2m}\frac{d^2}{dx^2} + V(x)\right)\psi(x) = E\psi(x) \qquad (1.18)$$

in *one dimension* with potential $V(x) = 0$ and particle mass $m = m_0$ has eigenfunctions $\psi(x)$ and energy eigenvalues $E = E_{\text{free}}$. The wave character of an electron in free-space is described by the wave function $\psi_k(x) = Ae^{i(kx-\omega t)}$ and the wave vector k is related to the eigenenergy by the dispersion relation

$$E_{\text{free}} = \frac{\hbar^2 k^2}{2m_0}. \qquad (1.19)$$

The solution to the Schrödinger equation describing an electron of energy E_{free} does not allow the particle to have any wave vector; rather, it requires the particle to have a wave vector k given by the dispersion relation, $E_{\text{free}}(k)$.

The *probability* of finding the particle described by a necessarily complex wave function $\psi_k(x) = Ae^{i(kx-\omega t)}$ at any position in space, $|\psi_k(x)|^2$, is *uniform*. No one location in space is more or less significant than any other location. The associated *continuous* translational invariance is a symmetry that, in a conservative system, gives rise to the conservation of linear momentum [2].

Since, according to equation (1.19), an electron of mass m_0 in free space has parabolic dispersion $E_{\text{free}} = \hbar^2 k^2/2m_0$, the second derivative of parabolic dispersion is inversely proportional to particle mass or, equivalently,

$$m_0 = \hbar^2 / \frac{d^2 E_{\text{free}}}{dk^2}. \qquad (1.20)$$

This shows that the inverse curvature of parabolic dispersion is proportional to particle mass. The momentum of the electron moving in one dimension in free space is $p = m_0 v_g = \hbar k$, the electron group velocity is $v_g = (\mathrm{d}E_{\text{free}}/\mathrm{d}k)/\hbar$, and the electron phase velocity is $v_p = E_{\text{free}}/\hbar k$.

1.4 Bloch's theorem

If a single electron is subject to a periodic potential $V(x)$ such that $V(x) = V(x + nL)$, where L is the minimum spatial period of the potential and n is an integer, then electron probability is modulated by the same periodicity as the potential. The isotropic electron probability symmetry of free space is broken and replaced by a new probability symmetry which is translationally invariant over a space spanned by a unit cell size of L. Electron probability is the same in each unit cell and is given by the function $|U_k(x)|^2$. It follows that the electron wave function is $U_k(x)$ to within a phase factor e^{ikx}. To find the phase factor, the eigenstates of the one-electron time-independent Schrödinger equation (equation (1.18)) must be solved. The electron wave function can be a *Bloch function* of the form

$$\psi_k(x) = U_k(x)e^{ikx}, \tag{1.21}$$

where $U_k(x) = U_k(x + nL)$ has the same periodicity as the potential $V(x) = V(x + nL)$. Electron probability $|\psi_k(x)|^2 = |U_k(x)|^2$ depends upon k but only contains the cell-periodic part of the wave function. The term e^{ikx} *modulates* the cell-periodic part of the Bloch function, carrying phase information between unit cells via the *Bloch wave vector k*. The dispersion relation $E(k)$ that relates the eigenenergy of the state $\psi_k(x)$ to the Bloch wave vector k is called the *band structure*.

The change in electron wave function moving from position x to $x + L$ is

$$\psi_k(x + L) = U_k(x + L)e^{ik(x+L)} = U_k(x)e^{ikx}e^{ikL} \tag{1.22}$$

$$\psi_k(x + L) = \psi_k(x)e^{ikL}. \tag{1.23}$$

Equation (1.23) is an expression of Bloch's theorem, which states that a potential with period L has wave functions separable into a part with the same period as the potential and a plane-wave term e^{ikL}.

Electron motion in a periodic potential involves the propagation of phase through the factor kx, where k is the Bloch wave vector. The Bloch wave function $\psi_k(x) = U_k(x)e^{ikx}$ is not an eigenfunction of the electron momentum operator, and so the value of *crystal momentum* $\hbar k$ is interpreted as an effective momentum of the electron. In a conservative system, the translational invariance of a periodic potential is a *discrete* symmetry that gives rise to the conservation of crystal momentum modulo $2\pi\hbar/L$. Hence, in *Umklapp scattering*, the final momentum after two particles collide in a periodic potential can differ from the total initial crystal momentum by modulo $2\pi\hbar/L$ with the lattice absorbing the excess momentum.

The Bloch wave function $\psi_k(x) = U_k(x)e^{ikx}$ is expressed in a plane-wave basis. It is also possible to use other basis functions, and this is considered next.

1.4.1 Wannier functions and Bloch's theorem

The Bloch condition given in equation (1.23) may also be expressed as a direct lattice sum of Wannier functions $\phi(x)$ or atomic orbitals localized around each lattice site $x_n = nL$, where n is an integer and L the lattice constant. The Wannier functions are related to the Bloch functions via the direct lattice sum over the N lattice sites of a finite-sized crystal:

$$\psi_k(x) = \frac{1}{\sqrt{N}} \sum_{n=1}^{n=N} e^{ikx_n} \phi(x - x_n) = \frac{1}{\sqrt{N}} \sum_n^N e^{iknL} \phi(x - nL). \tag{1.24}$$

Substituting $x = x + L$ gives

$$\psi_k(x + L) = \frac{1}{\sqrt{N}} \sum_n^N e^{iknL} \phi(x + L - nL) = \frac{1}{\sqrt{N}} \sum_n^N e^{iknL} \phi(x - (n-1)L). \tag{1.25}$$

Next, letting $n - 1 = m$ so that $n = m + 1$ and $x_m = mL$ gives

$$\psi_k(x + L) = \frac{1}{\sqrt{N}} \sum_m^N e^{ik(m+1)L} \phi(x - mL) = \frac{1}{\sqrt{N}} \sum_m^N e^{ikmL} \phi(x - mL) e^{ikL}$$

$$= \frac{1}{\sqrt{N}} \sum_m^N e^{ikx_m} \phi(x - x_m) e^{ikL}. \tag{1.26}$$

Making use of the expression for the direct lattice sum (equation (1.24)), equation (1.26) may be written as

$$\psi_k(x + L) = \psi_k(x) e^{ikL}, \tag{1.27}$$

which is just the Bloch condition, equation (1.23).

In the *delocalized* plane-wave Bloch functions $\psi_k(x) = U_k(x) e^{ikx}$, $U_k(x) = U_k(x + L)$ can be expressed as a direct lattice sum of *localized* states (Wannier functions) $\phi(x)$. The Wannier functions are localized around the lattice site x_n and are orthogonal for different lattice points so that

$$\int \phi^*(x - x_m) \phi(x - x_n) dx = \delta_{mn}. \tag{1.28}$$

The localized Wannier functions $\phi(x)$ are not restricted in the same way as the $U_k(x)$ of delocalized Bloch functions for which $U_k(x) = U_k(x + L)$.

1.5 The origin of complex band structure

The effect a static periodic potential has on noninteracting electrons with wave character can be illustrated by considering a one-dimensional array of rectangular potential barriers [3]. The solution to the Schrödinger equation is neatly summarized by a *complex band structure* consisting of dispersion, $E(k) = \hbar\omega(k)$, with *real* values of the Bloch wave vector k and dispersion with *complex* values of the Bloch wave vector k.

Figure 1.5 is a sketch of part of a periodic potential with a lattice constant L and a unit cell consisting of a rectangular barrier of thickness L_b, potential energy $V = V_0$,

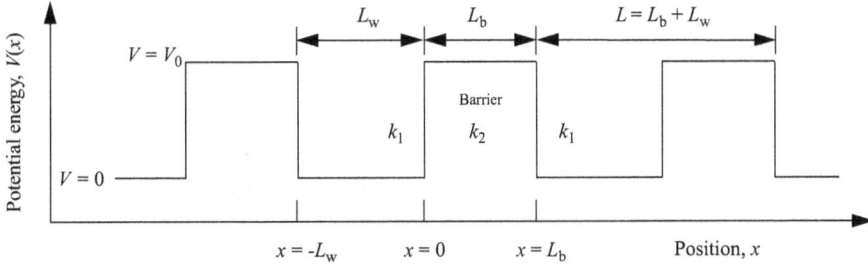

Figure 1.5. Periodic array of rectangular potential barriers with energy V_0 and thickness L_b. The potential wells have a thickness L_w, and the unit cell repeats in distance $L = L_b + L_w$ to form a lattice with lattice constant L. An electron of energy E has wave vector k_1 in the potential well and k_2 in the potential barrier.

a potential well of thickness L_w, and potential energy $V = 0$. If the potential well is labeled region 1 and the potential barrier is labeled region 2, the wave function for an electron of energy E in the unit cell is

$$\psi_1(-L_w \leqslant x < 0) = ae^{ik_1x} + be^{-ik_1x} \tag{1.29}$$

and

$$\psi_2(0 \leqslant x < L_b) = ce^{ik_2x} + de^{-ik_2x}. \tag{1.30}$$

For an electron of mass m_0 and energy $E > 0$, the value of k_1 in the potential well is

$$k_1 = \sqrt{2m_0E}/\hbar, \tag{1.31}$$

so that $E(k_1) = \hbar^2 k_1^2/2m_0$. The value of k_2 in the potential barrier region is

$$k_2 = \sqrt{2m_0(E - V_0)}/\hbar. \tag{1.32}$$

The amplitudes a, b, c, and d are found by applying the boundary conditions for the wave function that enforces continuity, smoothness, and the phase factor e^{ikL} required by Bloch's theorem. This gives

$$\psi_1(0) = \psi_2(0) \tag{1.33}$$

$$\frac{d\psi_1(0)}{dx} = \frac{d\psi_2(0)}{dx} \tag{1.34}$$

$$e^{ikL}\psi_1(-L_w) = \psi_2(L_b) \tag{1.35}$$

$$e^{ikL}\frac{d\psi_1(-L_w)}{dx} = \frac{d\psi_2(L_b)}{dx}. \tag{1.36}$$

Applying these conditions to equations (1.29) and (1.30) results in

$$a + b = c + d \tag{1.37}$$

$$(a - b)k_1 = (c - d)k_2 \tag{1.38}$$

$$e^{ikL}(ae^{-ik_1 L_w} + be^{ik_1 L_w}) = ce^{ik_2 L_b} + de^{-ik_2 L_b} \tag{1.39}$$

$$e^{ikL}(ae^{-ik_1 L_w} - be^{ik_1 L_w})k_1 = (ce^{ik_2 L_b} - de^{-ik_2 L_b})k_2. \tag{1.40}$$

These four linear homogeneous equations may be written in matrix form with a solution if the determinant is zero. That is,

$$\begin{vmatrix} 1 & 1 & -1 & -1 \\ k_1 & -k_1 & -k_2 & k_2 \\ -e^{ikL-ik_1 L_w} & -e^{ikL+ik_1 L_w} & e^{ik_2 L_b} & e^{-ik_2 L_b} \\ -k_1 e^{ikL-ik_1 L_w} & k_1 e^{ikL+ik_1 L_w} & k_2 e^{ik_2 L_b} & -k_2 e^{-ik_2 L_b} \end{vmatrix} = 0. \tag{1.41}$$

The solution gives rise to a *complex band structure* consisting of dispersion, $E(k) = \hbar\omega(k)$, with *real* values of the Bloch wave vector k and dispersion with *complex* values of the Bloch wave vector k. The corresponding real band structure describes *delocalized propagating electron states* in the crystal and the band structure with complex values of k can be used to describe *localized non-propagating electron states* [4]. Propagating electron states in the one-dimensional lattice have group velocity $v_g = d\omega/dk$ and the phase velocity is $v_p = \omega/k$.

The solution to the Schrödinger equation describing an electron of energy $E(k_1) = \hbar^2 k_1^2/2m_0$ in the periodic potential of figure 1.5 is the Bloch wave function given by equation (1.21). The particle is required to have a Bloch wave vector k given by the dispersion relation $E(k)$. In the one-dimensional periodic potential considered here, the propagating electron states at the conduction band minimum may be characterized using an effective electron mass $m_e^* = \hbar/(d^2\omega/dk^2)$.

1.5.1 The real electronic band structure

The solution to equation (1.41) for Bloch wave vector k is

$$\cos(kL) = \cos(k_2 L_b)\cos(k_1 L_w) - \frac{(k_2^2 + k_1^2)}{2k_1 k_2} \sin(k_2 L_b)\sin(k_1 L_w) = \theta_B, \tag{1.42}$$

so that the real band structure for *propagating* electron states of energy E and normalized *real* wave vector kL/π requires $|\theta_B| < 1$ and

$$kL = \frac{\mathrm{acos}(\theta_B)}{\pi}, \tag{1.43}$$

where θ_B is defined by equation (1.42). The energy dispersion relation $E(k)$ with real values of k, for which $|\theta_B| \leqslant 1$, is called the real band structure (or just band structure for short).

1.5.2 Complex electronic band structure

There are solutions to equation (1.42) in which energy dispersion has values of k that are complex-valued. In this case $|\theta_B| > 1$ and

$$kL = \frac{\mathrm{acosh}(\theta_B)}{\pi}. \tag{1.44}$$

The description of electron dispersion, which includes both real and complex values of k, is called the complex band structure.

Analytic continuation requires that the delocalized propagating ($|\theta_B| < 1$ with real k) and localized non-propagating ($|\theta_B| > 1$ with complex k) states in the dispersion relation connect smoothly. A delocalized non-propagating state with a real wave vector exists when $|\theta_B| = 1$. This corresponds to a standing wave at an extremum of the band, and so $dE/dk = 0$ such that electron group velocity is zero, $v_g = d\omega/dk = 0$. This is the point at which energy dispersion with the real wave vector connects to dispersion with the complex wave vector.

1.5.3 An example of complex band structure

Consider an electron of bare mass m_0 and energy E constrained to motion in the x-direction propagating in a static periodic potential consisting of a single rectangular barrier per unit cell that is illustrated in figure 1.5. In this particular case, the potential energy of the barrier is $V_0 = 3$ eV, the lattice constant is $L = 1$ nm, the potential barrier thickness $L_b = 0.4$ nm, and the potential well thickness $L_w = 0.6$ nm. Allowing the electron energy to have values in the range $0 \leqslant E_k \leqslant 4.6$ eV and plotting the real and imaginary contributions to the complex band structure in the reduced zone $0 \leqslant k < \pi/L$ gives the results presented in figure 1.6. The figure shows the existence of electron energy bands and band gaps.

The dispersion relation in each energy band *gap* is a function $E(k) = \hbar\omega(k)$ in which wave vector k has an imaginary component, $\mathrm{Im}(k) = \kappa$ with κ real. It follows that noninteracting electrons cannot occupy states with an energy eigenvalue that falls within the band gap energy of a perfect bulk crystal's static periodic potential. This is because solutions with $|\theta_B| > 1$ are of the form $e^{\mp\kappa x}$. These evanescent electronic states grow or decay exponentially from one unit cell to the next, are unbounded (diverge) at $x \to \mp\infty$, and are excluded by translational symmetry in a bulk crystal. Hence, noninteracting electrons in a static periodic potential cannot occupy band gap states. However, the usefulness of the solutions with $|\theta_B| > 1$ becomes apparent when translational symmetry is broken, which occurs in finite-sized semiconductor structures, tunnel barriers, heterointerfaces, lattice disorder, surfaces, and defects in nanoscale devices. In such situations, complex band structure can provide guidance and be an efficient way to anticipate and estimate exponential spatial variation of the electronic states that are the solution to the Schrödinger equation for a given nanoscale device configuration.

The value of band gap energy, E_g, decreases with increasing energy. Bands with real wave vectors, and hence propagating electron states, have energy bandwidths that increase with energy. Decreasing electron confinement in the potential wells as

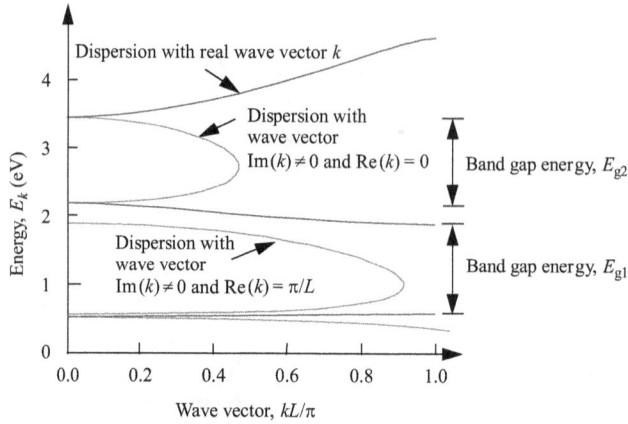

Figure 1.6. Calculated reduced-zone complex dispersion relation of an electron mass m_0, energy E, moving in a one-dimensional periodic potential consisting of rectangular potential barriers of energy $V_0 = 3$ eV, thickness $L_b = 0.4$ nm, and lattice constant $L = 1$ nm. The propagating states in a bulk crystal can be represented by a multivalued energy function of real wave vectors $k \in [0, \pi/L]$ (blue curves). Note that since $E(k) = E(-k)$ the Brillouin zone at $k = \pi/L$ is the same as at $k = -\pi/L$ and topologically a circle. In regions with an energy gap, E_g, wave vectors associated with non-propagating states in the bulk crystal take on complex values (red curves). The states in the energy gap E_{g1} have Im $(k) \neq 0$ and Re $(k) = \pi/L$. The states in the energy gap E_{g2} have Im $(k) \neq 0$ and Re $(k) = 0$. In general, the imaginary component of the wave vector is not bounded to the reduced-zone interval $[0, \pi/L]$ in an energy band gap and can have a value that exceeds π/L. The maximum magnitude of the imaginary component of the wave vector, |max Im(k)|, does not necessarily occur in the middle of the band gap.

electron energy increases has the effect of increasing the range in energy of the allowed propagating electron states.

As may be seen in figure 1.6, the maximum magnitude of the imaginary component of the wave vector $|\max \text{Im}(k)|$ in an energy band gap does *not* necessarily occur in the middle of the band gap.

For the case considered, the dispersion relation between electron energy E_k and Bloch wave vector k in a periodic potential has band extrema that may be approximated with parabolic dispersion. This gives rise to an effective mass, $m^* = m_{\text{eff}} \times m_0$, at a band extremum that is defined by the inverse curvature of the dispersion relation

$$m^* = \hbar^2 / \frac{\mathrm{d}^2 E_k}{\mathrm{d}k^2}. \tag{1.45}$$

1.6 The tight-binding method

The atoms that make up many semiconductors have outermost valence electrons that are s- or p-type and so can form s–p hybridized bonds. Examples from group IV of the periodic table and atoms contributing to III–V compound semiconductors include:

C	$[He]2s^22p^2$	Group IVA
Si	$[Ne]3s^23p^2$	Group IVA
Ge	$[Ar]3d^{10}4s^2p^2$	Group IVA
Ga	$[Ar]3d^{10}4s^2p^1$	Group IIIA
In	$[Kr]4d^{10}5s^2p^1$	Group IIIA
P	$[Ne]3s^23p^3$	Group VA
As	$[Ar]3d^{10}4s^2p^3$	Group VA

In crystal form, the atomic energy levels broaden into conduction and valence bands that retain some of the character of the s- and p-atomic states. The broadening of atomic energy levels into bands is illustrated in figure 1.7, which shows schematically what happens as a function of atomic separation. In a lattice characterized by nearest-neighbor atomic separation, L, energy bands and band gaps form.

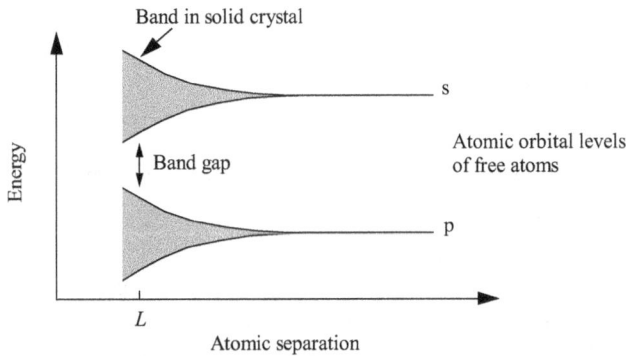

Figure 1.7. A schematic showing that interaction between atom electronic levels occurs as atomic separation decreases. For a large number of atoms in a lattice, this has the effect of creating a continuum of levels. In a lattice characterized by length L, energy bands and band gaps form.

The idea that the atomic character of electronic states is retained in band structure suggests that atomic states might form a good basis for constructing a physical model of the crystal. The tight-binding method does just that. Linear combinations of atomic orbitals (LCAOs) are used to calculate the band structure and the single-electron Bloch states of the crystal.

In the tight-binding method, little overlap between wave functions of atoms making up the crystal is assumed. This is useful because an expansion of overlap integrals in terms of neighbor separation can be truncated.

As described in section 1.5, the solutions to the one-electron Schrödinger equation for a potential consisting of an array of rectangular potential barriers include both propagating states with real k and non-propagating states with complex k. The same is true for electron states described by the tight-binding method. See appendix C for an introduction to tight-binding complex band structure.

1.7 Tight binding in three dimensions

In a bulk three-dimensional crystal there can be several basis functions per primitive unit cell. For example, the basis functions might include the s, p_x, p_y, p_z orbitals at each atomic site [5]. To accurately calculate band structure, an appropriate number, n_b, of basis functions must be chosen to represent the wave function, ϕ_n, in the nth primitive cell at each lattice position $\mathbf{R}_n$. Bloch's theorem requires

$$\phi_n = \phi_0 e^{i\mathbf{k}\cdot\mathbf{R}_n}, \tag{1.46}$$

where ϕ_n is a $n_b \times 1$ column vector, ϕ_0 is the wave function at the reference cell position, and $\mathbf{k}$ is the Bloch wave vector. In the nearest-neighbor tight-binding approximation, the Hamiltonian $\mathbf{H}$ is a square $n_b \times n_b$ matrix, and the solutions to

$$\sum_m \mathbf{H}_{0m}\phi_m = E(\mathbf{k})\phi_0 \tag{1.47}$$

have energy eigenvalues corresponding to the band structure, $E(\mathbf{k})$, for propagating electron states in the crystal.

1.7.1 The band structure of group IV and III–V semiconductors

The group IV semiconductor Si has a diamond crystal structure and the [Ne]$3s^2 3p^2$ atomic electronic structure. The [Ne] core is a closed shell, and four valence orbitals, 3s, $3p_x$, $3p_y$, and $3p_z$, are available to form bonds. In the sp^{3}s* model, an additional 4s basis function is included to improve the accuracy of the tight-binding band structure calculation. Since the diamond crystal structure has a basis of two atoms and in the sp^{3}s* model there are five electron orbitals per atom, in this case the number of basis functions is $n_b = 10$.

The III–V binary compound semiconductors, such as InAs, InP, and GaAs, have the zinc blende crystal structure with a two-atom basis, and their band structure can be obtained in a similar way.

This semi-empirical approach to calculating band structure requires estimating the matrix elements appearing in the Hamiltonian. Using the values given by Vogel [6], figure 1.8(a) shows the calculated band structure of GaAs in the important crystal symmetry directions.

Figure 1.8(b) is the GaAs band structure near the Γ-symmetry point. In a direct band gap semiconductor such as GaAs, the valence band maximum and the conduction band minimum occur at the same value of the Bloch wave vector. In a pure bulk semiconductor at the temperature $T = 0$ K, all electron states that form the crystal bonds are occupied in the valence band and none are occupied in the conduction band.

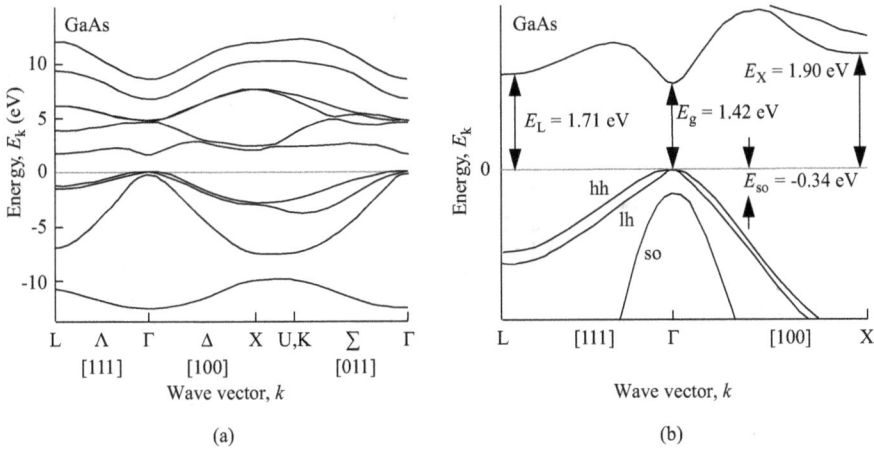

Figure 1.8. (a) The tight-binding band structure of GaAs calculated using the matrix element tabulation of Vogel [6]. (b) Detail of (a) near the Γ-symmetry point. The conduction band, heavy hole band (hh), light hole band (lh), and split-off band (so) are shown in the Γ-X [100] (Δ) and Γ-L [111] (Λ) crystal directions.

1.8 The semiconductor heterostructure

The operation of most electronic and optoelectronic components depends critically on the motion of charge carriers. Transistors switch, and lasers lase, because electrons carry current to and from the active region of the semiconductor device. The motion of charge carriers can be controlled using nanometer-scale lateral feature sizes, precisely defined heterostructures, and large electric fields. Light can be confined and controlled using heterostructure waveguides and materials patterned with nanoscale precision.

Since semiconductor crystals are composed of periodic arrays of atoms with electronic structures characterized by an energy band gap, it seems natural to parameterize semiconductors by their crystal lattice spacing, L, and band gap energy, E_g. Because the elemental semiconductors Si and Ge can be alloyed to form a crystal $Si_{1-\xi}Ge_\xi$ with Ge alloy fraction ξ, it is possible to plot their indirect band gap energy as a function of lattice spacing as a curve on an $E_g - L$ plot. This is shown as the thin red line in figure 1.9. Similarly, direct and indirect band gap binary III–V and II–VI compound semiconductors can form crystalline ternary and quaternary alloy semiconductors. The $E_g - L$ curves for some representative III–V and II–VI compound semiconductors with the zinc blende crystal structure are also shown in the figure.

In addition to band gap energy and crystal lattice spacing, other properties of semiconductors are important in determining device behavior. Table 1.1 lists some of these for a few compound semiconductors with the zinc blende crystal structure.

In table 1.1, the band gap energy of GaAs is $E_g = 1.424$ eV, the lattice constant is $L = 0.565325$ nm, the separation in energy between the conduction band at Γ and the subsidiary minimum in the L symmetry direction is $E_{\Gamma L} = 0.33$ eV, and between Γ and X it is $E_{\Gamma X} = 0.5$ eV. The heavy hole, light hole, and conduction band

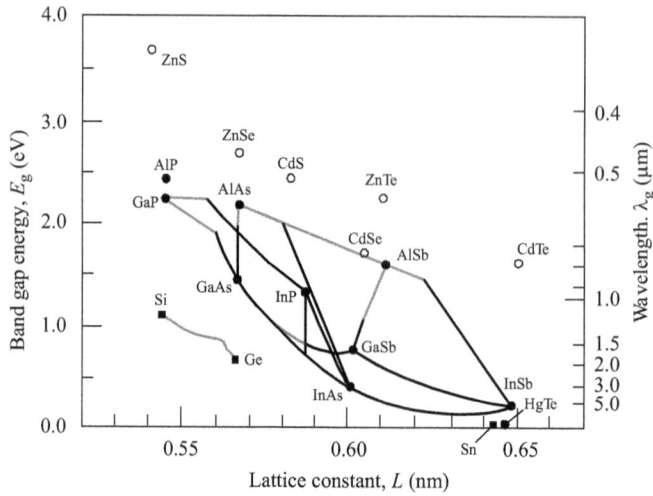

Figure 1.9. Location in the $E_g - L$ plane for the indicated semiconductors and their alloys. Black curves are semiconductors with a direct band gap and red curves are semiconductors with an indirect band gap.

Table 1.1. Room-temperature ($T = 300$ K) properties of direct band gap bulk single-crystal compound semiconductors with the zinc blende crystal structure.

	InSb	InAs	$In_{0.53}Ga_{0.47}As$	InP	GaSb	GaAs	GaN
E_g (eV)	0.17	0.354	0.75	1.344	0.75	1.424	3.2
L (nm)	0.647 937	0.605 83	0.5869	0.586 83	0.609 593	0.565 325	0.452
$E_{\Gamma L}$ (eV)	0.77	0.9	0.55	0.61	0.08	0.33	1.6
$E_{\Gamma X}$ (eV)	1.47	1.80	1.15	0.90	0.50	0.50	1.4
m_{hh}^*/m_0	0.44	0.41	0.5	0.64	0.33	0.5	1.3
m_{lh}^*/m_0	0.016	0.025	0.051	0.12	0.056	0.082	0.19
m_e^*/m_0	0.0145	0.021	0.042	0.077	0.048	0.067	0.13
ε_{r0}	17.9	14.55	13.88	12.35	15.69	13.1	9.6
$\varepsilon_{r\infty}$	15.7	11.8	11.34	9.52	14.44	11.1	5.35
$\hbar\omega_{LO}$ (meV)	24.4	30.2	33.5/31.5	42.7	29.8	36.3	92.1
$\hbar\omega_{TO}$ (meV)	22.9	27.1	30.2/28.0	38.2	26.8	33.3	68.8

effective electron mass at the Γ-point are $m_{hh}^* = 0.5 \times m_0$, $m_{lh}^* = 0.082 \times m_0$, and $m_e^* = 0.067 \times m_0$, respectively. The low-frequency relative permittivity is $\varepsilon_{r0} = 13.1$ and the high-frequency relative permittivity is $\varepsilon_{r\infty} = 11.1$. The longitudinal optical phonon energy is $\hbar\omega_{LO} = 36.3$ meV and the transverse optical phonon energy is $\hbar\omega_{TO} = 33.3$ meV. The polar-optic phonons in table 1.1 are related to relative permittivity via the Lyddane–Sachs–Teller relation, $(\omega_{LO}/\omega_{TO})^2 = \varepsilon_{r0}/\varepsilon_{r\infty}$.

Molecular beam epitaxy (MBE) or metalorganic chemical vapor deposition (MOCVD) can be used to grow single-crystal semiconductors with atomic layer precision on a planar or patterned substrate. In this way, an atomically abrupt interface between two semiconductors with different band gaps and similar lattice spacing can be created. Such single-crystal heterostructures have a conduction band minimum energy offset $\Delta E_{CB_{min}}$ and valence band maximum energy offset $\Delta E_{VB_{max}}$ at the heterointerface. The alignment in the energy of the conduction band and valence band edges, CB_{min} and VB_{max}, respectively, depends on the semiconductor material properties and a static dipole at the heterointerface. The dipole is formed by electrons tunneling from one semiconductor into the band gap of the other. This gap state model [7] necessarily involves the *complex band structure*. Since the complex band structure depends on the crystal symmetry direction, the planar heterostructure band alignment depends on the crystal orientation.

The three possible heterostructure band alignments are sketched in figure 1.10. They are the straddling gap (type-I), the staggered gap (type-II), and the broken gap (type-III).

An example of a type-I heterointerface occurs in the lattice-matched GaAs/$Al_\xi Ga_{1-\xi}As$ material system where ξ is the aluminum alloy fraction. At room temperature, the band gap energy of GaAs is smaller than $Al_\xi Ga_{1-\xi}As$ for which

$$E_g(Al_\xi Ga_{1-\xi}As) = 1.424 + (\xi \times 1.247) \text{ eV}. \tag{1.48}$$

The conduction band minimum energy offset at a planar GaAs/$Al_\xi Ga_{1-\xi}As$ heterointerface grown in the [001] crystal direction is

$$\Delta E_{CB_{min}} = \xi \times 0.84 \text{ eV}, \tag{1.49}$$

for $0 \leqslant \xi \leqslant 0.4$. Since ξ is a continuous variable, in this material system it is possible to tune the conduction band energy offset in the range $0 \leqslant \Delta E_{CB_{min}} \leqslant 0.34 \text{ eV}$ continuously.

A single-crystal planar heterostructure can be used to spatially control and localize electrons at the interface. A thin epitaxial planar layer of one semiconductor embedded in another can form a *quantum well* potential in which the electron confinement energy can be greater than energy broadening from electron scattering. Electron confinement and energy quantization in three dimensions can be achieved in small volumes, *quantum dots*, of one semiconductor embedded in another semiconductor.

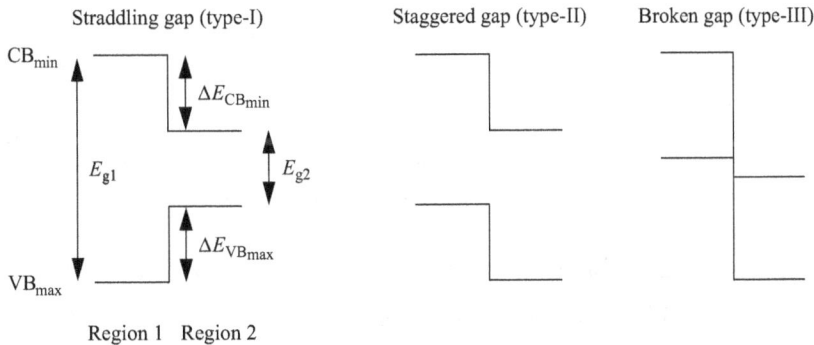

Figure 1.10. Sketch of band alignment at a heterointerface between two semiconductors with different band gap energies E_{g1} and E_{g2}.

Because single-crystal heterostructures can be grown with atomic layer precision, it is possible to engineer conduction and valence band potential profiles to *control* electron scattering and, in this way, determine current flow as a function of applied voltage bias in a device [8].

Other experimentally accessible control parameters include the strain resulting from a difference in lattice spacing between two epitaxially grown single-crystal semiconductors. Strain can usefully be used to modify and control band structure. However, there are limits to how much strain can be applied. Beyond a critical value, dislocations can form and cause severe degradation in device performance.

1.9 Substitutional doping of a semiconductor

Electrical conductivity in a semiconductor can be controlled via substitutional doping. Typically, a small number of dopant (impurity) atoms are introduced into a semiconductor crystal in such a way that it is energetically favorable for each impurity atom to replace an atom on a lattice site of the original crystal. For example, to introduce electrons (n-type carriers) into the conduction band of GaAs, a small number of Si donor impurity atoms can occupy Ga sites. At each impurity site, three of the four chemically active Si electrons replace the three Ga valence electrons, and the remaining Si electron can contribute to n-type electrical conduction. If the n-type doping concentration in the bulk semiconductor is $n = 10^{18}$ cm^{-3} then the average spacing between randomly positioned dopant atoms is 10 nm and, in GaAs, this results in good electrical conduction. The absence of electrons (p-type carriers, also called holes) can be introduced into the valence band of GaAs by doping with Be acceptor impurity atoms. In this case, it is energetically favorable for a Be atom to replace a Ga site. At each acceptor impurity site, the two chemically active Be electrons replace two of the three Ga valence electrons, and the missing electron can contribute to p-type electrical conduction.

Crystal growth techniques such as MBE and MOCVD may be used to grow and dope semiconductor crystals with atomic layer precision. By varying dopant concentration during epitaxial crystal growth, a thick planar layer of p-type semiconductor can be grown on a thick layer of n-type semiconductor to form a single-crystal p–n junction. Electron transport perpendicular to the junction interface has the electrical characteristics of a rectifying diode.

The same crystal growth method may be used to create a heterostructure diode in which both semiconductor material composition and doping are precisely controlled. For example, a single-crystal heterostructure consisting of an undoped (intrinsic) GaAs layer a few nm thick sandwiched between bulk p-type Al$_\xi$Ga$_{1-\xi}$As and bulk n-type Al$_\xi$Ga$_{1-\xi}$As with aluminum alloy fraction $\xi < 0.4$ forms a double heterostructure p–i–n junction that can be used to create a laser diode.

1.10 Thermal equilibrium and particle distribution functions

Thermodynamic equilibrium is characterized by an absolute temperature and may only exist when there is no net macroscopic flux of particles or energy within or between systems. In quantum mechanics, a large number (the thermodynamic limit) of identical indistinguishable noninteracting particles in thermal equilibrium are described by a state function with definite symmetry. The occupation probability of a particle in a state of energy E_k depends on particle spin. The electron is an elementary particle that has a spin of one-half, and the photon is an elementary particle that has a spin of one. A half-odd-integer-spin particle such as an electron is called a fermion. An integer-spin particle such as a photon with unity spin or phonon with a spin of zero is called a boson.

A large number of identical indistinguishable noninteracting half-odd-integer-spin particles of energy E_k in thermal equilibrium at absolute temperature T have a Fermi–Dirac probability distribution given by

$$f_k(E_k) = \frac{1}{e^{(E_k - \mu)/k_B T} + 1},$$ (1.50)

where k_B is the Boltzmann constant and μ is the chemical potential energy. The chemical potential appears in the distribution function because conservation of particles is assumed. In classical thermodynamics, the chemical potential is defined as the change in total energy needed to place an extra particle in a system of N particles at constant entropy S_{ent} and volume V_{vol} so that $\mu = (\partial E / \partial N)_{S_{ent}, V_{vol}}$. The chemical potential of a large number of noninteracting electrons in the low-temperature limit $T \rightarrow 0\ K$ is called the Fermi energy, $E_F = \mu(T \rightarrow 0\ K)$. If electrons of density n in a homogeneous isotropic bulk material are described by an electron mass m^*, then the Fermi wave number $k_F = (3\pi^2 n)^{1/3}$ and the value of the Fermi energy is $E_F = \hbar^2 k_F^2 / 2m^*$. At finite temperature and $E_k - \mu \gg k_B T$, equation (1.50) becomes the Maxwell–Boltzmann distribution, that is $e^{-E_k/k_B T}$.

The probability distribution for identical indistinguishable noninteracting integer-spin particles of energy $\hbar\omega$ in thermal equilibrium at temperature T is given by the Bose–Einstein function

$$g_{BE}(\hbar\omega) = \frac{1}{e^{(\hbar\omega - \mu)/k_B T} - 1}.$$ (1.51)

Usually, integer-spin particles such as phonons and photons are not conserved quantities. In this case $\mu = 0$ and the Bose–Einstein distribution becomes the Bose distribution

$$g_B(\hbar\omega) = \frac{1}{e^{\hbar\omega/k_B T} - 1}.$$ (1.52)

It is important to note that the operation of a conventional semiconductor laser diode is only possible by driving the system *out of thermal equilibrium*. A laser diode is driven out of thermal equilibrium by injection of an external current, I_{inj}, containing a net macroscopic flux of electrons and lasing emission removes a macroscopic flux of photons from the device. The transition from non-lasing to lasing emission in a macroscopic laser diode is an example of a *second-order nonequilibrium phase transition* in which the magnitude of the photon field is the order parameter.

1.11 Double heterostructure diode

A single-crystal p–i–n heterostructure diode can be designed to emit light efficiently. As illustrated in figure 1.11, the type-I heterointerface between a layered $GaAs/Al_\xi Ga_{1-\xi}$ structure with $\xi < 0.4$ can be used to separate doped p- and n-type $Al_\xi Ga_{1-\xi}As$ semiconductor from an undoped intrinsic optically active direct band gap GaAs gain region. A typical GaAs layer thickness is 100 nm, and the thickness of the adjacent $Al_\xi Ga_{1-\xi}As$ layers can be more than 1000 nm. Figure 1.11(a) shows the conduction and valence band profile of a $GaAs/Al_\xi Ga_{1-\xi}$ double heterostructure p–i–n junction calculated in the depletion approximation with p- and n-type doping concentration of 10^{18} cm^{-3}. There is a constant electric field in the GaAs i-region of the device. The lower refractive index of the $Al_\xi Ga_{1-\xi}As$ alloy compared to GaAs can be used to guide light in an optical cavity. When under forward voltage bias, V_{bias}, a current, I_{inj}, flows, and electrons and holes are injected into the optically active GaAs region. Figure 1.11(b) shows the conduction and valence band profile of a $GaAs/Al_\xi Ga_{1-\xi}$ double heterostructure when a high density of electrons and holes are injected into the GaAs i-region of the p–i–n junction. At high injection currents, the carrier density can be large enough to screen the electric field in the GaAs i-region. In an ideal p–n junction injection current, I_{inj}, is related to V_{bias} via

$$I_{inj} = I_{sat} (e^{eV_{bias}/k_B T} - 1), \tag{1.53}$$

where I_{sat} is the saturation current and T is the absolute temperature of the diode. Recombination of carriers that have been injected into the GaAs active region can result in the spontaneous emission of photons, and it is possible to configure the device as a light-emitting diode (LED). Alternatively, if the carrier density is large

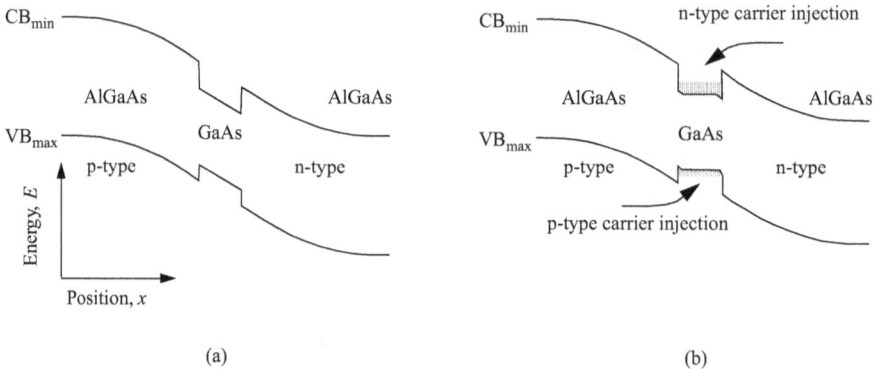

Figure 1.11. (a) Sketch illustrating conduction and valence band profile of a GaAs/AlGaAs double heterostructure p–i–n junction in the depletion approximation. The conduction band minimum (CB$_{min}$) and valence band maximum (VB$_{max}$) are shown in a cross-section through the GaAs/AlGaAs layer structure. (b) Sketch illustrating the conduction and valence band profile of a GaAs/AlGaAs double heterostructure p–i–n junction in the presence of n-carriers injected into the active region of the device. Notice that at high injected carrier density, the electric field in the GaAs can be screened.

enough, optical gain can result, and the device can be configured to operate as a laser diode (LD). Lasing occurs above a threshold value of injection current when optical gain can overcome optical losses. Replacing the GaAs active region with a few thin GaAs layers separated by $Al_\xi Ga_{1-\xi}As$ can form quantum wells whose carrier density for a given injection current can be large and result in reduced laser threshold current. The thickness of a typical quantum well is about 10 nm.

Bibliography

[1] Levi A F J 2023 *Applied Quantum Mechanics* 3rd edn (Cambridge: Cambridge University Press)
[2] Noether E 1918 *Math.-Phys. Kl.* **235**
[3] de R, Kronig L and Penney W J 1930 *Proc. R. Soc.* A **130** 499
[4] Under special circumstances, localized embedded states can exist in a continuum of delocalized states. For a review, see Hsu C W, Zhen B, Stone A D, Joannopoulos J D and Soljačić M 2016 *Nat. Rev. Mater.* **1** 16048
[5] Chadi D J and Cohen M L 1975 *Phys. Stat. Sol.* B **68** 405
[6] Vogl P, Hjalmarson H P and Dow J D 1983 *J. Phys. Chem. Solids* **44** 365
[7] Tersoff J 1984 *Phys. Rev.* B **30** 4874
[8] Levi A F J 2008 *Proc. IEEE* **96** 335

IOP Publishing

Essential Semiconductor Laser Device Physics (Second Edition)

A F J Levi

Chapter 2

Spontaneous emission and optical gain

This chapter develops the definitions of spontaneous emission, optical absorption, and optical gain in a direct band-gap semiconductor crystal. The relationship between optical absorption and spontaneous emission is described along with the Bernard–Duraffourg condition for optical gain. A thermodynamically consistent approach to calculating optical gain in the presence of electron scatting in an optical gain medium and the existence of Urbach band-edge spectral tails is provided, as are comments on the approach to more sophisticated models of spontaneous emission and optical gain in semiconductors.

Transitions between discrete bound electronic states of an atom can be driven (stimulated) by an external electric field and result in emission or absorption of a photon. An excited electron state can also spontaneously decay to a lower energy state via emission of a photon. There are selection rules for an atom to emit or absorb a photon via an oscillating dipole mechanism. In this case, transitions between electronic states of differing eigenenergy, with principal quantum numbers n' and n, respectively, require $l' - l = \Delta l = \pm 1$, in which the photon carries unit integer quantized angular momentum, $\hbar$. The change in quantum number $m' - m = \Delta m = \pm 1$, 0, in which 0 corresponds to linear photon polarization and with the sign convention $+ (-)$ for left (right)-handed circularly polarized photon absorption.

Unlike individual atoms, bulk semiconductors consist of many atoms, and there is a continuum of electronic energy levels in the conduction and valence bands. The presence of electrons and photons interacting in a semiconductor device driven to lasing can, in principle, be quite a challenge to model.

Spontaneous and stimulated emission play an essential role in the physics of semiconductor laser operation. Developing a description of spontaneous emission and optical gain that is both accurate and efficient is of practical importance because it enables semiconductor laser diodes to be designed optimally. In the following, a simplified model is developed that, remarkably, is accurate enough to predict laser

doi:10.1088/978-0-7503-6417-1ch2

diode behavior reliably. Sufficiently accurate models can be used to engineer and design efficient laser diodes optimally.

2.1 Spontaneous and stimulated emission

Following convention [1], consider a two-level electronic system. A schematic energy-level diagram for two states $|1\rangle$ and $|2\rangle$ showing stimulated and spontaneous processes is presented in figure 2.1(a). In the figure, $P(E)$ is the *spectral density* of electromagnetic modes measured in units of number of photons per unit volume per unit energy interval, and f_1 (f_2) is the ensemble average Fermi–Dirac occupation factor for electrons in state $|1\rangle$ ($|2\rangle$). It is assumed that the system is maintained at absolute temperature T.

An electromagnetic field can stimulate transitions between electronic states. Spontaneous transitions from an excited high-energy electronic state to a lower one can also occur. The existence of spontaneous emission is required to ensure that a system can reach thermal equilibrium. The stimulated emission coefficient between an occupied electronic state $|2\rangle$ and an unoccupied electronic state $|1\rangle$ is B_{21} and may be calculated using the golden rule. The corresponding stimulated emission rate is $B_{21}P(E_{21})f_2(1 - f_1)$. The spontaneous emission coefficient between states $|2\rangle$ and $|1\rangle$ is A_{21} and the corresponding spontaneous emission rate is $A_{21}f_2(1 - f_1)$.

In a conventional semiconductor laser diode, photon transitions occur between conduction and valence band states. The optically active region of a device where these transitions are important is typically a direct band-gap semiconductor, examples of which include GaAs, InP, and InGaAs. In such a semiconductor, the energy minimum of the conduction band lines up with the maximum energy of the valence band in k-space. This fact is of particular importance for the direct interaction of semiconductor electronic states with photons of wave vector k_{ph},

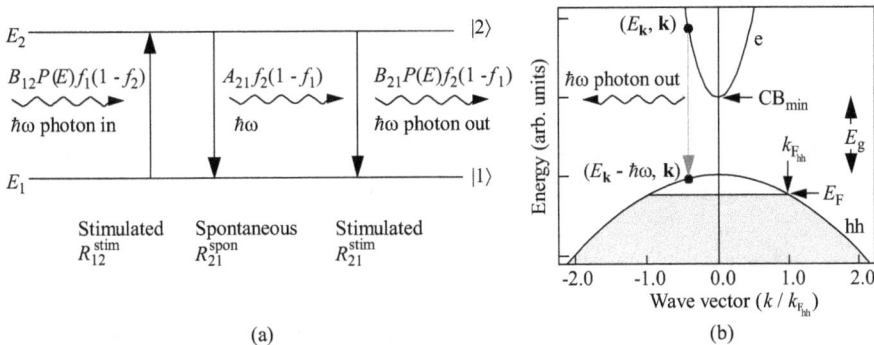

(a)   (b)

Figure 2.1. (a) Schematic energy-level diagram showing stimulated and spontaneous optical transitions between two electronic energy levels. (b) Band structure of a direct band-gap semiconductor showing the valence heavy-hole band hh, conduction band e, minimum conduction band energy CB_{min}, and band-gap energy E_g. The semiconductor is doped p-type and, at low temperatures, the Fermi energy is E_F and the Fermi wave vector is $k_{F_{hh}}$. Electrons in the conduction band can make a transition from a state characterized by the wave vector $\mathbf{k}$ and energy E_k in the conduction band to wave vector state $\mathbf{k}$ energy $E_k - \hbar\omega$ in the valence band by emitting a photon of energy $\hbar\omega$.

because the photon dispersion relation is typically linear and almost vertical compared with the dispersion relation for electrons in a given band extremum $\omega = \hbar k^2/2m^*$, where m^* is an effective mass. Conservation of momentum during a transition from occupied to unoccupied electronic states via the emission or absorption of a photon can, therefore, be approximated as a vertical transition in k-space.

The simplest model of a direct band-gap semiconductor typically includes a heavy-hole (hh) valence band with effective hole mass m_{hh}^* and conduction band (e) with effective electron mass m_e^*. Figure 2.1(b) shows schematically the spontaneous emission of a photon of energy $\hbar\omega$ accompanied by an electronic transition of a state characterized by wave vector $\mathbf{k}$ and energy $E_{\mathbf{k}}$ in the conduction band to wave vector state $\mathbf{k}$ energy $E_{\mathbf{k}} - \hbar\omega$ in the valence band. Because the initial and final electronic states have the same value of $\mathbf{k}$, it is assumed that negligible momentum is carried off by the photon.

If energy is measured from the top of the valence band, then the energy of an electron in the conduction band with effective electron mass m_e^* is

$$E_2 = E_g + \frac{\hbar^2 k^2}{2m_e^*} \tag{2.1}$$

and the energy of an electron in the valence band with effective electron mass m_{hh}^* is

$$E_1 = -\frac{\hbar^2 k^2}{2m_{hh}^*}. \tag{2.2}$$

The energy due to an electronic transition from the conduction band to the valence band is

$$\hbar\omega = E_2 - E_1 = \frac{\hbar^2 k^2}{2}\left(\frac{1}{m_e^*} + \frac{1}{m_{hh}^*}\right) + E_g. \tag{2.3}$$

A reduced effective electron mass m_r may be defined such that

$$\frac{1}{m_r^*} = \frac{1}{m_e^*} + \frac{1}{m_{hh}^*}. \tag{2.4}$$

For GaAs, appropriate values are $m_e^* = 0.07 \times m_0$ and $m_{hh}^* = 0.5 \times m_0$, giving $m_r^* = 0.06 \times m_0$. Equation (2.4) allows equation (2.3) to be written

$$\hbar\omega = \frac{\hbar^2 k^2}{2m_r^*} + E_g. \tag{2.5}$$

It follows that the three-dimensional density of electronic states coupled to vertical photon transitions of energy $\hbar\omega$ is

$$D_3(\hbar\omega) = \frac{1}{2\pi^2}\left(\frac{2m_r^*}{\hbar^2}\right)^{3/2}(\hbar\omega - E_g)^{1/2}. \tag{2.6}$$

If figure 2.1(b) represents the physical processes involved in an optically active semiconductor, then first-order time-dependent perturbation theory (the golden rule) might be used to calculate the transition probability between electronic states. In this case, all that needs to be known is the matrix element coupling the initial and final states and the density of the final states.

The photon modes of the system exist inside an optical cavity. As illustrated in figure 2.2, the cavity can be a cube of side L_{vol}, volume V_{vol}, that is in thermal equilibrium at absolute temperature T. For *periodic* boundary conditions, the allowed photon modes are k-states $k_{n_j} = 2\pi n_j/L_{\text{vol}}$, where $j = x, y, z$ in Cartesian coordinates and n is an integer. The spectral density $P(E)$ of photon modes at a specific energy is found by multiplying the density of photon states $D_3^{\text{ph}}(\hbar\omega)$ by the Bose occupation factor for photons

$$g_B(E) = \frac{1}{e^{E/k_B T} - 1}. \tag{2.7}$$

Since

$$D_3^{\text{ph}}(k_{\text{ph}})dk = \frac{1}{V_{\text{vol}}} 2\left(\frac{L_{\text{vol}}}{2\pi}\right)^3 4\pi k_{\text{ph}}^2 dk_{\text{ph}} = \left(\frac{k_{\text{ph}}}{\pi}\right)^2 dk_{\text{ph}} \tag{2.8}$$

and noting that $E = \hbar\omega = \hbar c k_{\text{ph}}$ in free space and $dE = \hbar c \, dk_{\text{ph}}$, results in

$$D_3^{\text{ph}}(E) = \frac{E^2}{\pi^2 \hbar^3 c^3} \tag{2.9}$$

giving a *spectral density* measured in units of the number of photons per unit volume per unit energy interval:

$$P(E) = D_3^{\text{ph}}(E)g_B(E) = \frac{E^2}{\pi^2 \hbar^3 c^3} \frac{1}{e^{E/k_B T} - 1}. \tag{2.10}$$

If the photon modes exist in a homogeneous isotropic dielectric medium characterized by refractive index n_r at frequency ω, then $E = \hbar\omega = \hbar c k_{\text{ph}}/n_r$. Further, if $n_r = n_r(\omega)$, then $dk_{\text{ph}} = (1/c)(n_r + \omega(dn_r/d\omega))d\omega$. Simplifying, by

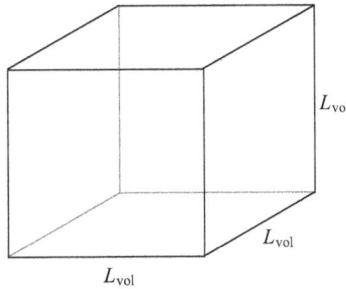

Figure 2.2. The spectral density of the photon field with periodic boundary conditions in a homogeneous isotropic cube of side L_{vol} in thermal equilibrium at temperature T is $P(E) = D_3^{\text{ph}}(E)g_B(E)$, where $D_3^{\text{ph}}(E)$ is the density of photon modes and $g_B(E)$ is the photon occupation factor.

ignoring dispersion in the refractive index ($\omega(\mathrm{d}n_{\mathrm{r}}/\mathrm{d}\omega) = 0$), the spectral density in the medium is

$$P(E) = \frac{E^2 n_{\mathrm{r}}^2}{\pi^2 \hbar^3 c^3} \frac{1}{e^{E/k_B T} - 1}. \tag{2.11}$$

Assuming the golden rule (the first term in the Born series for time-dependent perturbation theory) may be used to calculate the transition rates between states $|1\rangle$ and $|2\rangle$ of an atom inside an isotropic dielectric, it is natural to define a stimulated emission coefficient

$$B_{21} \equiv \frac{2\pi}{\hbar} |W_{21}|^2, \tag{2.12}$$

where $|W_{21}|^2$ is the magnitude of the matrix element squared coupling the initial and final states. The stimulated and spontaneous rates for photons of energy $E_{21} = E_2 - E_1$ become

$$R_{12}^{\text{stim}} = B_{12} P(E_{21}) f_1 (1 - f_2) \tag{2.13}$$

$$R_{21}^{\text{stim}} = B_{21} P(E_{21}) f_2 (1 - f_1) \tag{2.14}$$

$$R_{21}^{\text{spon}} = A_{21} f_2 (1 - f_1), \tag{2.15}$$

where the Fermi–Dirac distribution function gives the probability of non-interacting identical indistinguishable electron occupation

$$f_1 = \frac{1}{e^{(E_1 - \mu_1)/k_B T} + 1} \tag{2.16}$$

at energy E_1. A quasi-chemical potential level μ_1 is assumed, and the system is in thermal equilibrium characterized by temperature T. The probability of an unoccupied electron state at energy E_2 is $(1 - f_2)$.

Each electron particle has spin of one half and the ensemble average of non-interacting identical indistinguishable electrons in thermal equilibrium have a Fermi–Dirac distribution. In contrast, an *electron–hole pair excitation* (creation of an exciton) consists of an electron of spin one-half and an absence of an electron (hole) of spin one-half which, when combined, has integer spin and obeys a Bose–Einstein distribution.

When the two-level system illustrated in figure 2.1(a) is in thermal equilibrium, the net transition rates up and down in energy must balance, and there is only one quasi-chemical potential level (*the* chemical potential), so $\mu_1 = \mu_2 = \mu$. If thermal equilibrium exists between the two-level system and the photon modes of the cavity, then

$$R_{12}^{\text{stim}} = R_{21}^{\text{stim}} + R_{21}^{\text{spon}} \tag{2.17}$$

and so, making use of equations (2.13), (2.14), and (2.15),

$$B_{12} P(E_{21}) f_1 (1 - f_2) = B_{21} P(E_{21}) f_2 (1 - f_1) + A_{21} f_2 (1 - f_1), \tag{2.18}$$

which may be rearranged to give

$$P(E_{21})\big(B_{12}f_1(1 - f_2) - B_{21}f_2(1 - f_1)\big) = A_{21}f_2(1 - f_1), \tag{2.19}$$

so that

$$
\begin{aligned}
P(E_{21}) &= \frac{A_{21}f_2(1 - f_1)}{B_{12}f_1(1 - f_2) - B_{21}f_2(1 - f_1)} = \frac{A_{21}(f_2 - f_1 f_2)}{B_{12}(f_1 - f_1 f_2) - B_{21}(f_2 - f_1 f_2)} \\[2mm]
&= \frac{A_{21}\left(\frac{1}{f_1} - 1\right)}{B_{12}\left(\frac{1}{f_2} - 1\right) - B_{21}\left(\frac{1}{f_1} - 1\right)} = \frac{A_{21}}{B_{12}\left(\frac{1/f_2 - 1}{1/f_1 - 1}\right) - B_{21}} \\[2mm]
&= \frac{A_{21}}{B_{12}e^{E_{21}/k_B T}\left(\frac{e^{-\mu_2/k_B T}}{e^{-\mu_1/k_B T}}\right) - B_{21}}.
\end{aligned}
\tag{2.20}
$$

Since the system is in equilibrium, $\mu_1 = \mu_2$ and making use of equations (2.10) and (2.11),

$$P(E_{21}) = \frac{A_{21}}{B_{12}e^{E_{21}/k_B T} - B_{21}} = D_3^{\text{ph}}(E_{21})\frac{1}{e^{E_{21}/k_B T} - 1} \tag{2.21}$$

so

$$A_{21}(e^{E_{21}/k_B T} - 1) = D_3^{\text{ph}}(E_{21})(B_{12}e^{E_{21}/k_B T} - B_{21}). \tag{2.22}$$

Because this relationship must hold for *any* temperature when the system is in equilibrium, it follows that (for $T \to \infty$)

$$B_{12} = B_{21} \tag{2.23}$$

and separating out the temperature-dependent terms in equation (2.22) gives

$$\frac{A_{21}}{B_{12}} = \frac{E_{21}^2 n_r^3}{\pi^2 \hbar^3 c^3} = D_3^{\text{ph}}(E_{21}). \tag{2.24}$$

Equations (2.23) and (2.24) are the *Einstein relations*, which connect the stimulated and spontaneous emission coefficients.

The Einstein relations are derived for an electron system in thermal equilibrium with the photon modes of the cavity. The complete system may be characterized by a single temperature, T. An ensemble of identical two-level atoms is assumed to have non-interacting identical indistinguishable electrons characterized by a Fermi–Dirac occupation probability distribution. In particular, the effect of correlations between electrons due to the Coulomb interaction has been ignored. In addition, correlations due to coupling between electromagnetic modes and the electron states are assumed to be weak.

2.1.1 Spontaneous emission

In a semiconductor optical gain medium electron scattering times are typically short compared to the spontaneous emission time, and carriers relax to a quasi-equilibrium described by the Fermi–Dirac distribution, $f_{\mathbf{k}}$. When injected with carrier density (or excitation density), n is small (and, or, temperature, T, is high) such that the chemical potential $\mu \ll E_F$, where E_F is the Fermi energy, then $f_{\mathbf{k}}$ may be approximated by a Maxwell–Boltzmann distribution. To show this, consider the Fermi–Dirac distribution for band b. It is

$$f_{b,\mathbf{k}} = \frac{1}{e^{(E_{b,\mathbf{k}} - \mu_b)/k_B T} + 1}, \tag{2.25}$$

where the chemical potential is μ_b, and electron energy $E_b = \frac{\hbar^2 k^2}{2m_b^*}$ for a band with effective electron mass m_b^*. If the chemical potential is large and negative, then $E_{b,\mathbf{k}} - \mu_b \gg k_B T$ and the exponential term in $f_{b,\mathbf{k}}$ dominates the 1 in the denominator, and $f_{b,\mathbf{k}} \sim e^{\mu_b/k_B T} e^{-E_{b,\mathbf{k}}/k_B T}$ is the Maxwell–Boltzmann distribution.

Assuming a reduced electron mass m_r^* such that

$$\frac{1}{m_r^*} = \frac{1}{m_e^*} + \frac{1}{m_{hh}^*}, \tag{2.26}$$

where m_e^* is the conduction band effective electron mass and m_{hh}^* is the heavy-hole effective mass, it is possible to show analytically that the *total* photon spontaneous emission rate is proportional to n^2 and has temperature dependence $1/T^{3/2}$.

Letting

$$a_b = \frac{\hbar^2}{2k_B T m_b^*} \tag{2.27}$$

the carrier density in the band may be written as

$$
\begin{aligned}
n_b &= 4\pi \int_0^\infty 2 f_b k^2 \frac{dk}{(2\pi)^3} = \frac{e^{\mu_b/k_B T}}{\pi^2} \int_0^\infty e^{-a_b k^2} k^2 dk \\
&= -\frac{e^{\mu_b/k_B T}}{\pi^2} \frac{\partial}{\partial a_b} \left(\int_0^\infty e^{-a_b k^2} dk \right),
\end{aligned}
\tag{2.28}
$$

so that

$$n_b = -\frac{e^{\mu_b/k_B T}}{\pi^2} \frac{\sqrt{\pi}}{2} \frac{\partial}{\partial a_b} \frac{1}{\sqrt{a_b}} = \frac{e^{\mu_b/k_B T}}{\pi^2} \frac{\sqrt{\pi}}{2} \frac{a_b^{-3/2}}{2} = \frac{1}{4} \left(\frac{2 m_b^* k_B T}{\pi \hbar^2} \right)^{3/2} e^{\mu_b/k_B T}, \tag{2.29}$$

where use has been made of the standard integral

$$\int_0^\infty e^{-ax^2} dx = \frac{1}{2}\sqrt{\frac{\pi}{a}}. \tag{2.30}$$

The quantity to be calculated for *total* spontaneous emission may be written in a simplified form as

$$r_{\text{spon-total}} = \int r_{\text{spon}}(\hbar\omega)\mathrm{d}\omega = r_0 \int D_3(\hbar\omega) f_e\, f_{\text{hh}}\, \mathrm{d}\omega = r_0 \sum_k f_{e,\,\mathbf{k}} f_{\text{hh},\mathbf{k}}, \qquad (2.31)$$

where the sum is over all vertical electronic transitions in k-space. Setting

$$a_r = \frac{\hbar^2}{2k_B T m_r^*} \qquad (2.32)$$

and calculating the sum of all vertical transitions,

$$\sum_k f_{e,\mathbf{k}}\, f_{\text{hh},\mathbf{k}} = \frac{e^{\mu_e/k_B T} e^{\mu_{\text{hh}}/k_B T}}{\pi^2} \int_0^\infty e^{-a_r k^2} k^2 \mathrm{d}k$$

$$= -\frac{e^{\mu_e/k_B T} e^{\mu_{\text{hh}}/k_B T}}{\pi^2} \frac{\partial}{\partial a_r} \int_0^\infty e^{-a_r k^2} \mathrm{d}k \qquad (2.33)$$

so that

$$\sum_k f_{e,\mathbf{k}} f_{\text{hh},\mathbf{k}} = -\frac{e^{\mu_e/k_B T} e^{\mu_{\text{hh}}/k_B T}}{\pi^2} \frac{\sqrt{\pi}}{2} \frac{\partial}{\partial a_r} \frac{1}{\sqrt{a_r}} = e^{\mu_e/k_B T} e^{\mu_{\text{hh}}/k_B T} \frac{1}{4}\left(\frac{2m_r^* k_B T}{\pi \hbar^2}\right)^{3/2}. \qquad (2.34)$$

Further, since from equation (2.29),

$$e^{\mu_b/k_B T} = 4 n_b \left(\frac{\pi \hbar^2}{2 m_b^* k_B T}\right)^{3/2} \qquad (2.35)$$

and because charge neutrality requires $n_e = n_{\text{hh}} = n$, equation (2.34) may be written as

$$\sum_k f_{e,\mathbf{k}} f_{\text{hh},\mathbf{k}} = n^2 4 \left(\frac{\pi \hbar^2 m_r^*}{2 m_e m_{\text{hh}} k_B T}\right)^{3/2}, \qquad (2.36)$$

from which it is clear that, in this limit, total spontaneous emission is proportional to n^2 and decreases with increasing temperature according to $1/T^{3/2}$. For large carrier densities (typically $n > 10^{18}\,\text{cm}^{-3}$ at room temperature) spontaneous emission increases at a rate *less* than n^2 because the constraint $f_{b,\mathbf{k}} \leqslant 1$ starts to become important.

2.1.2 Absorption and its relation to spontaneous emission

To establish a relationship between absorption and spontaneous emission, it is simplest to consider the case when the system is driven from equilibrium such that $\mu_1 \neq \mu_2$ but bath temperature, T, remains the same.

Photons of energy $E = \hbar\omega$ incident on a two-level system can cause transitions between two states $|1\rangle$ and $|2\rangle$ with energy eigenvalues $E_1 < E_2$. Average optical absorption in the medium, α_{opt}, may be defined as the ratio of the average number of absorbed photons per second per unit volume to the average number of incident photons per second per unit area. Hence,

$$\alpha_{\text{opt}} = \frac{R_{\text{net}}^{\text{stim}}}{S_{\text{Poynting}}/\hbar\omega} = \frac{R_{12}^{\text{stim}} - R_{21}^{\text{stim}}}{S_{\text{Poynting}}/\hbar\omega}, \tag{2.37}$$

where S_{Poynting} is the magnitude of the Poynting vector and $S_{\text{Poynting}}/\hbar\omega$ is the average number of incident photons per second per unit area. It is usual for α_{opt} to be measured in units of cm^{-1}. Since the absorption coefficient times the photon flux is the net stimulated rate,

$$\alpha_{\text{opt}} = \frac{B_{12} P(E_{21}) f_1 (1 - f_2) - B_{21} P(E_{21}) f_2 (1 - f_1)}{P(E_{21}) \frac{c}{n_r}}. \tag{2.38}$$

The ratio of spontaneous emission and absorption is

$$\frac{R_{21}^{\text{spon}}}{\alpha_{\text{opt}}} = \frac{A_{21} f_2 (1 - f_1)}{\frac{n_r}{c} B_{12} (f_1 - f_2)} = \frac{A_{21}(1 - f_1)}{\frac{n_r}{c} B_{12}\left(\frac{f_1}{f_2} - 1\right)} = \frac{A_{21}\left(\frac{1}{f_1} - 1\right)}{\frac{n_r}{c} B_{12}\left(\frac{1}{f_2} - \frac{1}{f_1}\right)}$$

$$= \frac{A_{21} e^{(E_1 - \mu_1)/k_B T}}{\frac{n_r}{c}} = \frac{A_{21}}{\frac{n_r}{c}}. \tag{2.39}$$

Substituting for the ratio A_{21}/B_{12} using equation (2.24) gives

$$\frac{R_{21}^{\text{spon}}}{\alpha_{\text{opt}}} = \frac{c}{n_r} D_3^{\text{ph}}(E_{21}) \frac{1}{e^{E_{21}/k_B T} e^{-(\mu_2 - \mu_1)/k_B T} - 1} = \frac{E_{21}^2 n_r^2}{\pi^2 c^2 \hbar^3} \frac{1}{e^{(E_{21} - \Delta\mu)/k_B T} - 1}, \tag{2.40}$$

where $\Delta\mu = \mu_2 - \mu_1$ is the difference in quasi-chemical potential levels used to describe the distribution of electronic states at energy E_2 and E_1, respectively. The approximation made is that equation (2.24) (which is derived for the equilibrium condition $\Delta\mu = 0$) remains valid when $\Delta\mu \neq 0$. This approximation is most likely to be appropriate when $\Delta\mu < k_B T$. The relationship between absorption and spontaneous emission for a system characterized by absolute temperature T and difference in chemical potential $\Delta\mu$ given by equation (2.40) may be rewritten in a convenient form as

$$\alpha_{\text{opt}} = \frac{\pi^2 c^2 \hbar^3}{E_{21}^2 n_r^2} R_{21}^{\text{spon}} (e^{(E_{21} - \Delta\mu)/k_B T} - 1). \tag{2.41}$$

Net optical gain exists when absorption α_{opt} is negative. Since spontaneous emission is always positive, the only way the value of absorption α can change sign is if the right-hand side term in parenthesis in equation (2.41) changes sign. This may be seen by rewriting equation (2.39) as

$$\alpha_{\text{opt}} = \frac{\frac{n_r}{c} R_{21}^{\text{spon}} B_{12} (f_1 - f_2)}{A_{21} f_2 (1 - f_1)}. \tag{2.42}$$

The denominator in equation (2.42) is always positive, since $0 < f_1 < 1$ and $0 < f_2 < 1$. The numerator is positive, giving positive absorption α_{opt} if $f_1 > f_2$ and the numerator is negative, giving negative absorption (or optical gain $g_{opt} \equiv -\alpha_{opt}$) if $f_1 < f_2$. The condition for optical gain is $f_2 - f_1 > 0$, or

$$\Delta\mu > E_{21}. \tag{2.43}$$

Hence, the separation in quasi-chemical potential levels must be greater than the photon energy for net optical gain to exist. Equation (2.43) is called the Bernard–Duraffourg condition [2].

In a bulk semiconductor, there are not just two energy levels E_2 and E_1 to be considered, but rather a continuum of energy levels in the conduction band and valence band. This is illustrated in figure 2.3. Electrons in the conduction band have a Fermi–Dirac distribution f_2, and in the valence band they have a distribution f_1. For equal carrier concentrations in the conduction band and valence band at fixed temperature, the Fermi–Dirac distribution functions f_2 and f_1 are different since, in general, the quasi-chemical potential levels are different. This occurs because the effective electron mass in each band is different, giving a different density of states, and hence different quasi-chemical potential levels for a given carrier concentration n and absolute temperature T.

For convenience, the energy of holes (absence of electrons) in the valence band can be measured from the valence band maximum *down* so that the energy of holes is negative, and the distribution of holes is $f_{hh} = (1 - f_1)$. The energy of electrons in the conduction band can be measured from the conduction band minimum *up* so that the energy of electrons is positive and the distribution of electrons is $f_e = f_2$. When calculating photon energy, $\hbar\omega$, for an inter-band transition it is necessary to

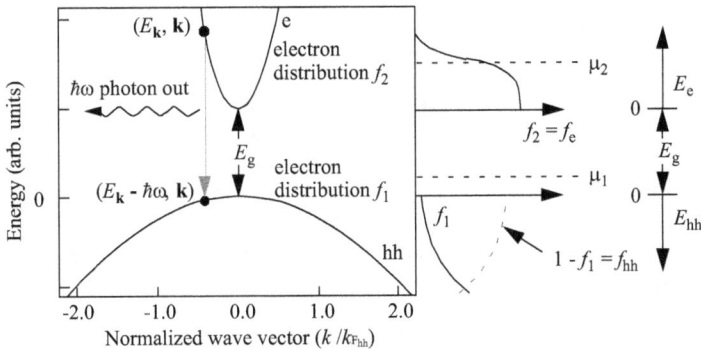

Figure 2.3. A schematic diagram showing an electronic transition from a state of energy E_k and wave vector k to a state of energy $E_k - \hbar\omega$ and wave vector **k**, resulting in the emission of a photon of energy $\hbar\omega$. A conduction band electron has an effective electron mass m_e^* and a valence band electron has an effective electron mass m_{hh}^*. The Fermi–Dirac distribution of electron states in the conduction band is f_2, and in the valence band it is f_1. The quasi-chemical potential levels are μ_2 and μ_1, respectively. Sometimes, it is convenient to measure electron energy E_e from the conduction band minimum and hole energy E_{hh} from the valence band maximum.

add the band-gap energy, E_g. Also, the condition for optical gain previously given by equation (2.43) becomes

$$\Delta\mu_{e-hh} > 0. \tag{2.44}$$

For GaAs, an appropriate value for electron mass is $m_e^* = 0.07 \times m_0$ and effective hole mass is $m_{hh}^* = 0.5 \times m_0$. If temperature $T = 300$ K and carrier density is fixed in value at $n = 1 \times 10^{18}$ cm^{-3} in each band, then $\mu_{hh} = -55$ meV and $\mu_e = 39$ meV (the value of μ_e is μ_2 with energy E_g subtracted). In this situation, $\Delta\mu_{e-hh}$ $=\mu_2 - \mu_1 - E_g = \mu_e + \mu_{hh} = -16$ meV and, according to equation (2.44), GaAs is optically absorbing for all photon energies $\hbar\omega > E_g$. On the other hand, when $n = 2 \times 10^{18}$ cm^{-3} in each band and $T = 300$ K, the chemical potentials are $\mu_{hh} = -36$ meV and $\mu_e = 75$ meV. Now $\Delta\mu_{e-hh} = 39$ meV and optical gain exists for photon energies, $\hbar\omega$, such that $E_g < \hbar\omega < E_g + \Delta\mu_{e-hh}$.

The total spontaneous emission $r_{spon}(\hbar\omega)$ for photons of energy $E = \hbar\omega$ is the sum of all energy levels separated by vertical transitions of energy E. Substituting equation (2.24) into equation (2.15), using the definition of $|W_{21}|^2$ given by equation (2.12), and performing the sum over allowed vertical k-state transitions results in

$$r_{spon}(\hbar\omega) = \frac{2\pi}{\hbar} \frac{E_{21} n_r^3}{\pi^2 \hbar^3 c^3} \sum_{k_1, k_2} |W_{21}|^2 f_2 (1 - f_1) \delta(E_{21} - \hbar\omega), \tag{2.45}$$

where the delta function ensures energy conservation. If the matrix element W_{21} is slowly varying as a function of E_{21}, it may be treated as a constant. Converting the sum to an integral and substituting $f_e = f_2$ and $f_{hh} = (1 - f_1)$ gives

$$r_{spon}(\hbar\omega) = \frac{2\pi}{\hbar} \frac{E_{21}^2 n_r^3}{\pi^2 \hbar^3 c^3} |W_{21}|^2 \int \frac{d^3 k}{(2\pi)^3} f_e f_h \delta\left(E_g + \frac{\hbar^2 k^2}{2m_r^*} - \hbar\omega\right), \tag{2.46}$$

which may be written as

$$r_{spon}(\hbar\omega) = \frac{2\pi}{\hbar} \frac{\hbar^2 \omega^2 n_r^3}{\pi^2 \hbar^3 c^3} |W_{21}|^2 \frac{1}{2\pi^2} \left(\frac{2m_r^*}{\hbar^2}\right) (\hbar\omega - E_g)^{1/2} f_e f_{hh}. \tag{2.47}$$

In this expression, $(\hbar\omega - E_g)^{1/2}$ is the energy dependence of the reduced three-dimensional density of electronic states in equation (2.6).

It follows from equation (2.41) that at equilibrium optical gain $g_{opt}(\hbar\omega)$ is related to $r_{spon}(\hbar\omega)$ through

$$g_{opt}(\hbar\omega) = -\alpha_{opt}(\hbar\omega) = \hbar \left(\frac{c\pi}{n_r \omega}\right)^2 r_{spon}(\hbar\omega)(1 - e^{(\hbar\omega - E_g - \Delta\mu_{e-hh})/k_B T}). \tag{2.48}$$

If $\Delta\mu_{e-hh} \neq 0$ then the electron and hole system is out of thermal equilibrium, even though the same temperature is assigned to the electron and hole subsystems. In a strict sense, the critical assumption of equilibrium used in deriving equation (2.48) is invalid. This has consequences such as, for example, the refractive index n_r depending upon $\Delta\mu_{e-hh}$.

An expression for optical gain may also be found directly by substituting equation (2.12) into equation (2.38) and performing the integral over the allowed initial and final electronic density of states (equation (2.6)). This gives

$$g_{opt}(\hbar\omega) = \frac{2\pi n_r}{c\hbar} \, |W_{21}|^2 \frac{1}{2\pi^2} \left(\frac{2m_r^*}{\hbar^2}\right)^{3/2} (\hbar\omega - E_g)^{1/2} (f_e + f_{hh} - 1). \qquad (2.49)$$

2.2 Optical transitions using the golden rule

The matrix element W_{21} appearing in equations (2.47) and (2.49) remain to be evaluated. This may be done by applying the golden rule.

Consider a semiconductor illuminated with light. The interaction between a classical optical electric field of the form

$$\mathbf{E}_{opt} = \mathbf{E}_0 e^{i(\mathbf{k}_{opt} \cdot \mathbf{r} - \omega t)} \qquad (2.50)$$

and an electron can be described in the dipole approximation by the perturbation [3]

$$\hat{W} = -e \, (\mathbf{r} \cdot \mathbf{E}_{opt}) \qquad (2.51)$$

Equation (2.51) is a semiclassical approximation because $\mathbf{E}_{opt}$ is the classical optical electric field.

Electron quantum mechanical states in a crystal are Bloch functions and so the dipole matrix element coupling a conduction band initial-state $\psi_e(\mathbf{r}) = U_{e,\mathbf{k}}(\mathbf{r})e^{i\mathbf{k}\cdot\mathbf{r}}$ and a heavy-hole valence band final state $\psi_{hh}(\mathbf{r}) = U_{hh,\mathbf{k}'}(\mathbf{r})e^{-i\mathbf{k}'\cdot\mathbf{r}}$ is

$$W_{ehh} = \langle\psi_{hh} | \hat{W} | \psi_e\rangle = -e \int U^*_{hh,\mathbf{k}'}(\mathbf{r}) U_{e,\mathbf{k}}(\mathbf{r}) \, \mathbf{r} \cdot \mathbf{E}_0 \, e^{-i(\mathbf{k}'-\mathbf{k}-\mathbf{k}_{opt})\cdot\mathbf{r}} \, d^3\mathbf{r} \qquad (2.52)$$

The term $e^{-i(\mathbf{k}'-\mathbf{k}-\mathbf{k}_{opt})\cdot\mathbf{r}}$ in the integral oscillates rapidly, resulting in $W_{ehh} \to 0$ *except* when $\mathbf{k}' - \mathbf{k} = \mathbf{k}_{opt}$. Since electronic states have $|\mathbf{k}| \sim 10^8 \, \text{cm}^{-1}$ and $|\mathbf{k}_{opt}| \sim 10^5 \, \text{cm}^{-1}$, it is reasonable to set $|\mathbf{k}_{opt}| = 0$, so that $\mathbf{k}' = \mathbf{k}$. As described in section 2.1, transitions between initial and final electron states conserve crystal momentum and so have essentially the same k-vector. This is why direct optical transitions between electronic states in the band structure are called vertical transitions.

Finding the value of the matrix element in equation (2.52) requires detailed knowledge of the Bloch wave functions involved in the transition. The calculations can be quite complicated [3]. To simplify matters, the coefficients, including the magnitude of the matrix element squared in equations (2.47) and (2.49), may be approximated as constants. The $\hbar^2\omega^2$ term in equation (2.47) is slowly varying and may be treated as a constant since the energy range of approximately $\Delta\mu_{e-hh}$ around

E_g is important and typically $\Delta\mu_{e-hh}/E_g \ll 1$. This allows the spontaneous emission (equation (2.47)) to be written as

$$r_{\text{spon}}(\hbar\omega) = r_0\left(\hbar\omega - E_g\right)^{1/2} f_e f_{hh}, \tag{2.53}$$

where r_0 is a material-dependent constant. Optical gain becomes

$$g_{\text{opt}}(\hbar\omega) = g_0\left(\hbar\omega - E_g\right)^{1/2}\left(f_e + f_{hh} - 1\right), \tag{2.54}$$

where the coefficient g_0 is also a material-dependent constant. The ratio $g_0/r_0 = \pi^2 c^2/\omega^2 n_r^2$. In this simple model, the range in energy over which optical gain exists is given by the difference in chemical potential, $\Delta\mu_{e-hh}$.

Obviously, the use of constants r_0 and g_0 results in quite a crude approximation. However, it does allow the estimation of trends, such as the temperature dependence of gain. In fact, it turns out that creating a model of optical gain in a semiconductor that is qualitatively more advanced is a very challenging task and the subject of ongoing research.

Figure 2.4 shows the result of calculating $g_{\text{opt}}(\hbar\omega)$ using equation (2.54) for the indicated carrier densities, band-gap energy $E_g = 1.4$ eV, temperature $T = 300$ K, effective electron mass $m_e^* = 0.07 \times m_0$, effective hole mass $m_{hh}^* = 0.5 \times m_0$, and $g_0 = 2.64 \times 10^4$ cm^{-1} eV$^{-1/2}$. The parameters used are appropriate for the direct band-gap semiconductor GaAs.

According to the results shown in figure 2.4, there is no optical gain when injected carrier density $n = 1 \times 10^{18}$ cm^{-3}. As carrier density increases, optical gain first appears near the band-gap energy, $E_g = 1.4$ eV. When carrier density $n = 2 \times 10^{18}$ cm^{-3}, a peak optical gain of 330 cm^{-1} occurs for photon energy near $\hbar\omega = 1.415$ eV, and the gain bandwidth is $\Delta\mu_{e-hh} = 39$ meV.

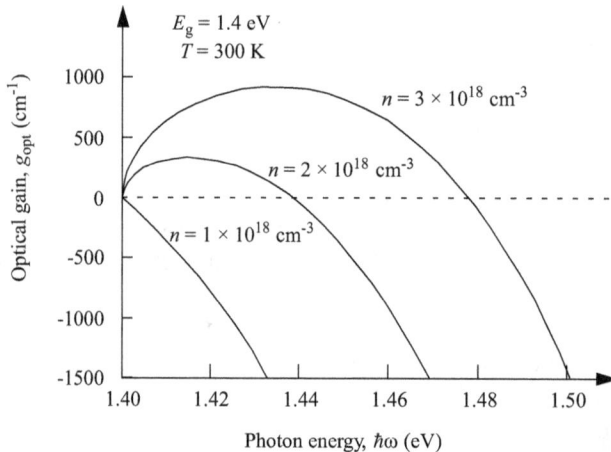

Figure 2.4. Calculated optical gain for indicated carrier densities, n, band-gap energy $E_g = 1.4$ eV, absolute temperature $T = 300$ K, effective electron mass $m_e^* = 0.07 \times m_0$, effective hole mass $m_{hh}^* = 0.5 \times m_0$, and $g_0 = 2.64 \times 10^4$ cm^{-1} eV$^{-1/2}$.

2.2.1 Optical gain in the presence of electron scattering

Inelastic electron scattering has the effect of energy-broadening electron states. Because a typical inelastic electron scattering rate $1/\tau_{in}$ can be tens of ps^{-1}, corresponding to several meV broadening, this effect is significant and on the same scale as the difference in chemical potential. *If* the probability of an electron state scattering may be approximated as $1 - e^{-t/\tau_{in}}$ then it has a Lorentzian broadened spectrum. Such a broadening function has an energy full-width-at-half-maximum $\gamma_k = \hbar/\tau_{in}$, so that if, for example, $\tau_{in} = 25$ fs then $\gamma_k = 26$ meV. There is a subscript k in γ_k because, in general, the scattering rate depends upon electron crystal momentum $\hbar k$. However, in practice, this fact is usually ignored, and γ_k is treated as a constant $\gamma_k = \gamma_0$.

To calculate the optical gain in the presence of electron scattering characterized by a single scattering rate, spontaneous emission is evaluated first using the Lorentzian broadening function to simulate the effect of electron–electron scattering. Equation (2.53) is modified to

$$r_{\text{spon}}(\hbar\omega) = r_0 \int_0^\infty E^{1/2} f_e \, f_{\text{hh}} \frac{\gamma_0/2\pi}{(E_g + E - \hbar\omega)^2 + \left(\dfrac{\gamma_0}{2}\right)^2} dE. \tag{2.55}$$

Optical gain g_{opt} as a function of photon energy $\hbar\omega$ is then calculated using equation (2.48). This ensures that optical transparency in the semiconductor occurs at a photon energy of $\Delta\mu_{e-hh} + E_g$, where $\Delta\mu_{e-hh}$ is the difference in chemical potential and E_g is the band-gap energy. Optical transparency occurring at a different energy violates the laws of thermodynamics.

Unfortunately, even this elementary consideration is often ignored in conventional theories, which put Lorentzian broadening directly in the gain function (equation (2.49)) [4]. As illustrated in figure 2.5, it is straightforward to show that not

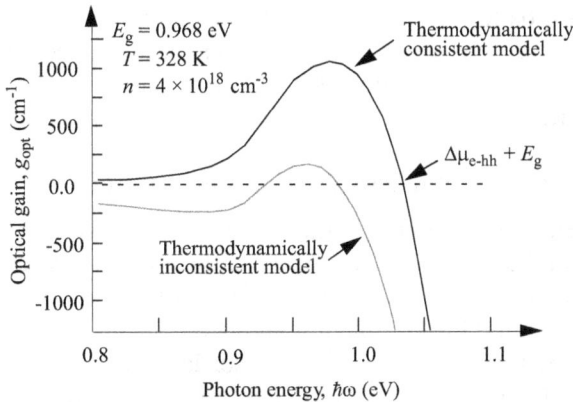

Figure 2.5. Calculated optical modal gain, including effects of electron–electron scattering in bulk InGaAsP with room-temperature band-gap energy $E_g = 0.968$ eV and an inelastic scattering broadening factor $\gamma_0 = 25$ meV.

only does this result in optical transparency at an energy less than $\Delta\mu_{e-hh} + E_g$, but it also predicts substantial absorption of sub-band-gap energy photons [5]—something that is not observed experimentally.

It is important to be able to correctly calculate gain spectra used in models of semiconductor laser diodes to ensure designs can be properly optimized. Any model should be able to accurately predict both peak optical gain and optical gain bandwidth as a function of injected carrier density. The fact that such details can matter is well illustrated by again noting that a relaxation-time approximation for electron scattering results, via a Fourier transform, in Lorentzian spectral broadening of electronic levels whose value is on the same scale as the gain bandwidth. Simple exponential Markovian decay of an electronic state with time dependence $e^{-t/\tau_{in}}$ dominated by a *single* time constant $\tau_{in} = \hbar/\gamma_0$ has spectrally long tails that are *not* present in measured gain spectra. An ad hoc replacement of the Lorentzian function in equation (2.55) with a sech function

$$\frac{1}{\pi\gamma_0} \, \text{sech} \left(\frac{E - \hbar\omega}{\gamma_0} \right) \tag{2.56}$$

suppresses long spectral tails and allows a better fit to experimentally measured spectra. Nevertheless, placing the broadening function given by equation (2.56) directly into the gain function (equation (2.49)) does not remove the inconsistency in which optical transparency is predicted to occur at an energy less than $\Delta\mu_{e-hh} + E_g$.

Any self-consistent description of band-edge spectral tails (*Urbach* tails) should take into account energy-dependent electron scattering and correlations of physical processes in the *complex* band structure. Urbach tails can be due to disorder, the presence of localized states, electron scattering, and many-body effects, all of which may break the discrete translational symmetry experienced by non-interacting electrons in the idealized static period potential of a semiconductor crystal. Electron scattering near the band edge can involve localized states with energy in the band gap and propagating states with energy in the band. Adopting a too simplistic model of optical gain near the band edge of a semiconductor can result in inaccurate predictions and hinder optimal device design. On the other hand, unnecessarily complicated models are not guaranteed accurate while also being computationally expensive thereby making them less useful for semiconductor device optimization algorithms.

2.3 Comments on the success of a simple model

The physics of electron scattering and its contribution to electron transport in semiconductors can be quite involved [6]. It is, therefore, truly remarkable that the relatively simple model of optical gain summarized by equation (2.48) gives useful results. Engineers can use this and slightly improved versions of the model to design and optimize the performance of semiconductor laser diodes successfully. The fact that this is so is largely accidental. If a more detailed description of physical processes in a device is desired it is difficult to make practical progress.

For example, the low-temperature energy dependence of the low carrier density optical absorption in bulk GaAs is very different from the simple square-root behavior predicted by equation (2.54) and based on a model that assumes non-interacting electrons. If Coulomb interactions are included, optical absorption can occur via other mechanisms. The translational symmetry that excludes non-interacting electrons from occupying states in the energy band gap of a bulk semiconductor can be broken by interactions. This may be illustrated by considering the Coulomb interaction attracting a negatively charged electron in the conduction band to a positively charged hole in the valence band and forming a localized correlated state called an exciton [7]. Absorption due to excitons can give rise to a spectrally sharp absorption peak for photons with energy less than the semiconductor band gap. At room temperature, this peak is broadened by scattering processes but still contributes to the absorption of photon energies near the band-gap energy. Excitonic absorption strength can be enhanced in quantum wells and controlled by applying an electric field perpendicular to the plane of the quantum well. This quantum-confined Stark effect [8] can be used to design optical modulators and detectors that operate at room temperature [9–11].

For carrier densities at values necessary to achieve useful optical gain in a laser diode, electron scattering broadens electron energy levels in the semiconductor on an energy scale that is comparable to optical gain bandwidth, $\Delta \mu_{e-hh}$. In addition, the relatively high carrier density screens the Coulomb interaction. This reduces the ability of the system to form excitons. High carrier density can also reduce the value of the band-gap energy.

While these and other high carrier density effects make it very hard to create a detailed physical model of optical gain in a semiconductor, it is also responsible for the success of a naive approach that assumes non-interacting electrons. Associated with high carrier density and room-temperature operation are energy-broadening effects that quite accidentally and serendipitously allow a simple model to be useful. The fact remains, however, that while the model gives results that may be used to design lasers, the model itself is physically incorrect.

Relying too heavily on a crude model can result in misunderstanding and misinterpreting device behavior, so care must be taken not to draw incorrect conclusions. There is a trade-off between using a simplified model of device behavior that is *good enough* for the target application and developing a more accurate but computationally more expensive description of physical behavior.

A portion of this chapter was reproduced with permission from [12].

Bibliography

[1] For example Casey H C and Panish M B 1978 *Heterostructure Lasers, Part A: Fundamental Principles* (Orlando, FL: Academic)

[2] Bernard M G A and Duraffourg G 1961 *Phys. Status Solidi* **1** 699

[3] For an introduction, see Chuang S L 1995 *Physics of Optoelectronic Devices* (New York: Wiley)

[4] For example, 1993 *Quantum Well Lasers* ed P Zory (New York: Academic) ; Contributions from Corzine, Yang, Coldren, Asada, Kapon, Englemann, Shieh, and Shu all explicitly and incorrectly put Lorentzian broadening directly in the gain function. Also, figure 2.4 in Chow W W and Koch S W 1999 *Semiconductor-Laser Fundamentals: Physics of the Gain Materials* (Berlin: Springer)

[5] For example Chuang S L, O'Gorman J and Levi A F J 1993 *IEEE J. Quantum Electron.* **29** 1631

[6] Levi A F J 2020 *Essential Electron Transport for Device Physics* (Melville, NY: AIP Publishing)

[7] This is just one of several different excitations that contribute to optical processes in semiconductors. For a review of such phenomena, see Schäfer W and Wegener M 2002 *Semiconductor Optics and Transport Phenomena* (Berlin: Springer)
Haug H and Koch S W 2009 *Quantum Theory of the Optical and Electronic Properties of Semiconductors* (Singapore: World Scientific)

[8] Miller D A B, Chemla D S, Damen T C, Gossard A C, Wiegmann W, Wood T H and Burns C A 1984 *Phys. Rev. Lett.* **53** 2173

[9] Miller D A B, Chemla D S, Damen T C, Gossard A C, Wiegmann W, Wood T H and Burns C A 1984 *Appl. Phys. Lett.* **45** 13

[10] Huang X, Seeds A J, Roberts J S and Knights A P 1998 *IEEE Photonics Technol. Lett.* **10** 1697

[11] Thalken J, Li W, Haas S and Levi A F J 2004 *Appl. Phys. Lett.* **85** 121

[12] Levi A F J 2023 *Applied Quantum Mechanics* 3rd edn (Cambridge: Cambridge University Press)

IOP Publishing

Essential Semiconductor Laser Device Physics (Second Edition)

A F J Levi

Chapter 3

The semiconductor laser diode

This chapter presents a brief history of the laser diode and the essential elements to consider in the design of a laser diode. A summary of typical optical cavity geometries is provided. The longitudinal optical resonances of a Fabry–Perot cavity, photon cavity round-trip time, optical spectrum, optical mode spacing, and cavity Q are discussed, as well as the optical mode profile in an index-guided slab waveguide. The optical loss of a mirror, internal waveguide optical loss, and photon lifetime are presented, along with the essential aspects of buried heterostructure Fabry–Perot laser diode design.

Fundamentally, new concepts can take decades to evolve into practical technology. The semiconductor laser diode is just such an example, with an approximately 30 year development period from a conceptual proof-of-principle demonstration to a mature technology in volume production. The history of the laser dates back to at least 1951 and an idea by Townes, who wanted to use ammonia molecules to amplify microwave radiation. Townes and two students completed a prototype device in late 1953 and named it *maser* or *microwave amplification by stimulated emission of radiation*. In 1958, Townes and Schawlow published results of a study showing that a similar device could be made to amplify light. The device was named a *laser*, which is an acronym for *light amplification by stimulated emission of radiation*. In principle, a large flux of essentially single-wavelength electromagnetic radiation could be produced by a laser. Independently, Prokhorov and Basov proposed related ideas. The first laser used a rod of ruby and was constructed in 1960 by Maiman. In late 1962, lasing action was reported in a pulsed current-driven GaAs p–n diode maintained at liquid nitrogen temperature (77 K) [1]. The Fabry–Perot optical cavity used to stabilize lasing wavelength was created by reflectivity from the parallel end-faces of the semiconductor diode. This integration of an optical resonator with a semiconductor p–n diode was so effective and practical that by 1970 continuous room-temperature operation of AlGaAs/GaAs double

doi:10.1088/978-0-7503-6417-1ch3

heterostructure laser diodes was reported [2] using a similar geometry and other improvements soon followed.

The potential use of laser diodes in telephone communication systems was recognized early. However, it took some time before useful devices and suitable glass-fiber transmission media became available. The first fiber-optic telephone installation was put in place in 1977 and consisted of a 2.4 km long link under downtown Chicago. By the early 1980s, the distributed feedback (DFB) semiconductor laser diode was being used to transmit data long distances via communication links, each involving optical transmission over distances of up to 100 km in single-mode glass fiber.

Another type of heterostructure laser diode suitable for use in data communication applications was inspired by the work of Iga published in 1977 [3]. By the late 1990s, these vertical-cavity surface-emitting lasers (VCSELs) had appeared in volume-manufactured commercial products.

After the introduction of the compact disk (CD) in 1982 and the digital versatile disk (DVD) in 1997, large-volume production of Fabry–Perot heterostructure semiconductor laser diodes was associated with optical disk applications [4]. Today, vast numbers of laser diodes are volume-manufactured for fiber-optic communication products, laser printers, cell phones, and laser copiers. There is additional low-volume production of laser diodes for numerous specialty markets.

3.1 Designing a laser diode

The existence of optical gain may be exploited to make a laser diode out of a direct band-gap semiconductor such as GaAs or InGaAsP. When a single-crystal direct band-gap semiconductor heterostructure p–n diode is forward-biased to inject a current, I_{inj}, electrons are introduced into the conduction band and holes into the valence band. In the optically active region of a semiconductor, the wave functions of electrons and holes can overlap in real space, and this is where vertical optical transitions can take place in k-space. If the density of carriers injected into the active region is great enough, then the Bernard–Duraffourg condition $\Delta\mu_{e-hh} > 0$ is satisfied and optical gain exists for light at some wavelength in the semiconductor. There is, however, more to designing a useful device. Among other things, the p–n diode should have a low series resistance, the laser should be efficient at converting electrical current to light, operate over a practical range of temperatures, and be designed so that a high intensity of lasing light emission occurs at a specific wavelength. The choice of semiconductor material and design of the optical cavity has a direct impact on device performance.

Because a typical value of gain for an optical mode in a semiconductor laser diode is not very large ($\sim$500 cm^{-1}), and to precisely control emission wavelength, the active semiconductor is typically placed in a high-Q optical cavity. The optical cavity has the effect of storing light at a particular wavelength, allowing it to interact with the gain medium for a longer time. In this way, a relatively modest optical gain may be used to build up high light intensity in a given optical mode. Electrons contributing to injection current, I_{inj}, are converted into lasing photons that occupy a single mode. The efficiency of this conversion process is enhanced if only one high-Q optical cavity resonance is in the same wavelength range as the semiconductor optical gain.

3.1.1 The optical cavity

Figures 3.1(a)–(c) illustrate optical cavities into which an optically active semi-conductor may be incorporated to form a Fabry–Perot laser, VCSEL [5], and microdisk laser [6], respectively.

Other types of optical cavities that are commonly used are the distributed feedback (DFB) and distributed Bragg grating (DBR) structures [7]. Hybrid devices consisting of an optical gain region and a physically separate passive high-Q optical resonator have also been explored. A well-established example is an extended cavity

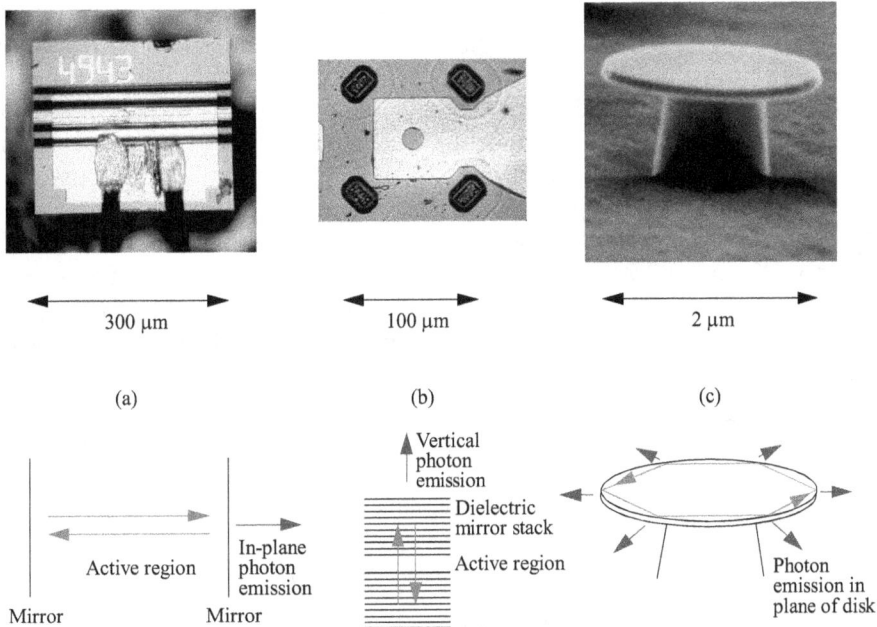

Figure 3.1. (a) Photograph of the top view of a Fabry–Perot, edge-emitting, semiconductor laser diode showing the horizontal gold metal stripe used to make electrical contact with the p-type contact of the diode. The n-type contact is made via the substrate. Two gold wire bonds attach to the large gold pad in the lower half of the picture. The device has a multiple-quantum-well InGaAsP active region, lasing emission at 1310 nm wavelength, a room-temperature laser threshold current of $I_{th} = 3$ mA, and a diode series resistance of 3 Ω. The sketch shows the side view of the 300 μm-long optical cavity formed by reflection at the cleaved semiconductor-air interface. (b) Photograph of the top view of a VCSEL showing gold metalization used to make electrical contact to the p-type contact of the diode. Lasing light is emitted from the small aperture in the center of the device. This VCSEL has a multiple-quantum-well GaAs active region, lasing emission at 850 nm wavelength, and a room-temperature laser threshold current of $I_{th} = 1$ mA. (c) Scanning electron microscope image of a microdisk laser. The semiconductor disk is 2 μm in diameter and 0.1 μm thick. The device has a single quantum-well InGaAs active region, lasing emission at 1550 nm wavelength, and a room-temperature external incident optical laser threshold pump power at a 980 nm wavelength of 300 μW.

formed by coupling a laser diode active region to a single-mode glass fiber with a Bragg grating [8]. Another hybrid device couples a semiconductor optical gain medium to Si_3N_4 ring resonators implemented using a silicon photonics platform to create a Kerr optical frequency comb source [9]. More speculative configurations that can be incorporated into cavity design include bound states in the continuum [10] and topologically protected edge states [11].

Because the Fabry–Perot resonator involves concepts that may be explored with relative ease, this is considered next.

3.1.2 Fabry–Perot longitudinal resonances

The Fabry–Perot semiconductor laser diode consists of an index-guided active-gain region placed within a Fabry–Perot optical resonator. In the Fabry–Perot optical cavity, photons propagate in the z-direction, normal to the two mirror planes, to form longitudinal optical resonances. *Index guiding* helps maintain the z-oriented trajectory of photons traveling perpendicular to the mirror plane. Index guiding is achieved in a buried heterostructure laser diode by surrounding the semiconductor active region with a semiconductor of lower refractive index. Usually, this involves etching the semiconductor wafer to define a narrow, z-oriented active-region stripe and then planarizing the etched regions by epitaxial growth of nonactive, lower refractive index, wider band-gap semiconductor.

Optical loss for a photon inside the Fabry–Perot cavity is minimized at cavity resonances. Figure 3.2 shows a schematic diagram of a Fabry–Perot optical resonator consisting of a semiconductor active-gain medium and two mirrors with reflectivity r_1 and r_2, respectively, forming an optical cavity of length L_C. The photon round-trip time in this cavity is τ_{RT}.

If mirror reflectivity is such that $r_1 = r_2 = 1$ then the Fabry–Perot cavity has an optical mode spacing given by $kL_C = \pi m$, where $m = 1, 2, 3, \ldots$, and $k = \omega n_r/c$. The refractive index of the dielectric is n_r, and c is the speed of light in a vacuum. Adjacent modes are spaced in angular frequency according to

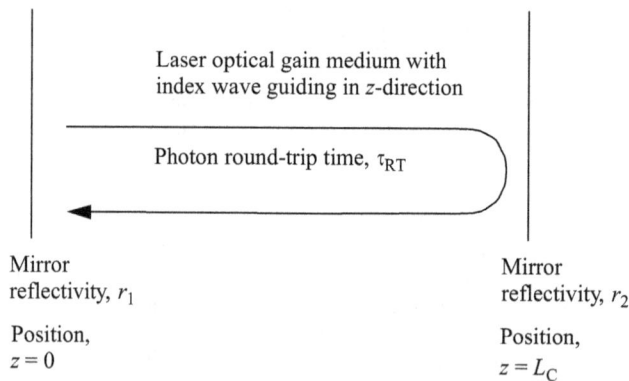

Laser optical gain medium with
index wave guiding in z-direction

Photon round-trip time, τ_{RT}

Mirror
reflectivity, r_1

Mirror
reflectivity, r_2

Position,
$z = 0$

Position,
$z = L_C$

Figure 3.2. Schematic diagram of a Fabry–Perot optical resonator consisting of a semiconductor active-gain medium and two mirrors with reflectivity r_1 and r_2, respectively, forming an optical cavity of length L_C.

$$\Delta\omega = \frac{c(k_{m+1} - k_m)}{n_r} = \frac{c\pi\,(m + 1 - m)}{L_C n_r} = \frac{c\pi}{L_C n_r} = 2\pi\Delta\nu. \tag{3.1}$$

This mode spacing is also called the free spectral range of the cavity. Measured in units of Hz,

$$\Delta\nu = \frac{c}{2L_C n_r} = \frac{1}{\tau_{RT}}, \tag{3.2}$$

where τ_{RT} is the round-trip time for a photon in the cavity and $\nu = \omega/2\pi$. The mode spacing as a function of wavelength is

$$\Delta\lambda = \frac{\lambda^2}{2L_C n_r}. \tag{3.3}$$

The spectral intensity as a function of frequency inside a Fabry–Perot cavity is [12]

$$I_{opt}(\nu) = \frac{I_{max}}{1 + \left(\frac{2\mathcal{F}}{\pi}\right)^2 \sin\left(\frac{\pi\nu}{\Delta\nu}\right)}, \tag{3.4}$$

where $\mathcal{F}$ is the finesse of the optical cavity

$$\mathcal{F} \equiv \frac{\pi\sqrt{r}}{1 - r} \tag{3.5}$$

and

$$I_{max} = \frac{I_0}{(1 - r)^2} \tag{3.6}$$

in which I_0 is the initial optical intensity in the cavity. The round-trip attenuation factor for light amplitude in the cavity is r. If the only optical loss is from the two mirrors with reflectivity r_1 and r_2, respectively, then $r = r_1 r_2$. For a lossless dielectric Fabry–Perot etalon with refractive index n_r, the mirror reflectivity at a cleaved dielectric-to-air interface is $r_{1,2} = |(1 - n_r)/(1 + n_r)|^2$.

The expression for $I_{opt}(\nu)$ may be derived by considering a Fabry–Perot optical resonator consisting of two parallel plane mirrors and cavity length L_C. The round-trip phase accumulated by a plane-wave electromagnetic field propagating in the z-direction, normal to the mirror, with wave vector $k = 2\pi/\lambda$ is

$$\phi = 2kL_C = \frac{4\pi}{\lambda}L_C = 4\pi L_C \frac{\nu}{c}. \tag{3.7}$$

On resonance, the electromagnetic field must accumulate an integer multiple of 2π. The condition is

$$k_0 L_C = n\pi, \tag{3.8}$$

where $n = 1, 2, 3, \ldots$, and k_0 is the resonant wave vector. The possibility of resonant zero or negative phase accumulation is usually ignored.

In general, if the initial value of the electric field amplitude is $\mathbf{E}_0$ and the *fractional loss* of amplitude remaining after one round trip in the cavity is $r < 1$, then, the electric field, after one round trip in the cavity, is

$$\mathbf{E}_1 = \mathbf{E}_0 r e^{-i\phi}. \tag{3.9}$$

If the electric field is allowed to propagate indefinitely in the resonator then the total electric field is an infinite sum of terms

$$\mathbf{E} = \mathbf{E}_0(1 + re^{-i\phi} + r^2 e^{-i2\phi} + r^3 e^{-i3\phi} + \ldots) = \frac{\mathbf{E}_0}{1 - re^{-i\phi}} \tag{3.10}$$

since

$$\sum_{n=0}^{n=\infty} ax^n = a\frac{1}{1 - x} \tag{3.11}$$

for $x < 1$ and constant a.

If only a finite number, N, of round-trips occur, then the total electric field is

$$\mathbf{E}_N = \mathbf{E}_0 \frac{1 - re^{-iN\phi}}{1 - re^{-i\phi}} \tag{3.12}$$

since

$$\sum_{n=0}^{n=N-1} ax^n = a\frac{1 - x^N}{1 - x}. \tag{3.13}$$

For the infinite sum case, the total electric field intensity in the resonator is

$$I_{\text{opt}} = |\mathbf{E}|^2 = \left| \frac{\mathbf{E}_0}{1 - re^{-i\phi}} \right|^2 = \frac{I_0}{(1 - r\cos(\phi))^2 + (r\sin(\phi))^2}$$

$$= \frac{I_0}{1 + r^2 - 2r\cos(\phi)}, \tag{3.14}$$

where intensity $I_0 = |\mathbf{E}_0|^2$. Adding and subtracting $2r$ in the denominator gives

$$1 + r^2 - 2r + 2r - 2r\cos(\phi) = (1 - r)^2 + 2r(1 - \cos(\phi)). \tag{3.15}$$

Notice that $2\sin(x)\sin(y) = \cos(x - y) - \cos(x + y)$, then when $x = y$ it follows that $2\sin^2(x) = 1 - \cos(2x)$ and so

$$1 + r^2 - 2r + 2r - 2r\cos(\phi) = (1 - r)^2 + 4r\left(1 - \sin^2\left(\frac{\phi}{2}\right)\right). \tag{3.16}$$

Hence, equation (3.14) may be written

$$I_{\text{opt}} = \frac{I_0}{(1 - r)^2 - 4r\sin^2\left(\frac{\phi}{2}\right)} \tag{3.17}$$

and making use of the definitions of free spectral range (equation (3.2)), finesse (equation (3.5)), and I_{max} (equation (3.6)) results in

$$I_{opt} = \frac{I_{max}}{1 + \left(\frac{2\mathcal{F}}{\pi}\right)^2 \sin^2\left(\frac{\phi}{2}\right)} = \frac{I_{max}}{1 + \left(\frac{2\mathcal{F}}{\pi}\right)^2 \sin^2\left(\frac{\pi\nu}{\Delta\nu}\right)}, \quad (3.18)$$

which agrees with equation (3.4).

When finesse is *large* ($\mathcal{F} \gg 1$), the optical linewidth γ_{opt} at the resonant frequency ν_0 is much smaller than the free spectral range $\Delta\nu$, so that $\mathcal{F} = \Delta\nu/\gamma_{opt}$. In this limit of $\mathcal{F} \gg 1$, the expression for full-width-half-maximum of the optical resonance becomes $\gamma_{opt} = \Delta\nu/\mathcal{F}$ and the optical-Q associated with the cavity is the frequency ν_0 of the resonance divided by γ_{opt}, and so $Q = \nu_0/\gamma_{opt}$. However, in general, Q is defined in terms of the ratio of real and imaginary frequency so that

$$Q \equiv \frac{1}{2}\left|\frac{\nu_{Re}}{\nu_{Im}}\right|. \quad (3.19)$$

Only when the imaginary part of frequency $\nu = \nu_{Re} + i\nu_{Im}$ is small, so that $\nu_{Im} = \gamma_{opt} \ll \nu_0$, does $Q \to \nu_0/\gamma_{opt}$.

A Fabry–Perot laser diode with cavity length $L_C = 300\,\mu$m, an effective refractive index $n_r = 3.3$, and emission wavelength near $\lambda = 1310$ nm, corresponding to a frequency $\nu_0 = 229$ THz, has optical resonator mode spacing $\Delta\nu = 151$ GHz or $\Delta\lambda = 0.867$ nm. The spectral intensity as a function of frequency, when $r_1 = r_2 = 0.286$, is shown in figure 3.3(a). Finesse is $\mathcal{F} = 0.978$. Also

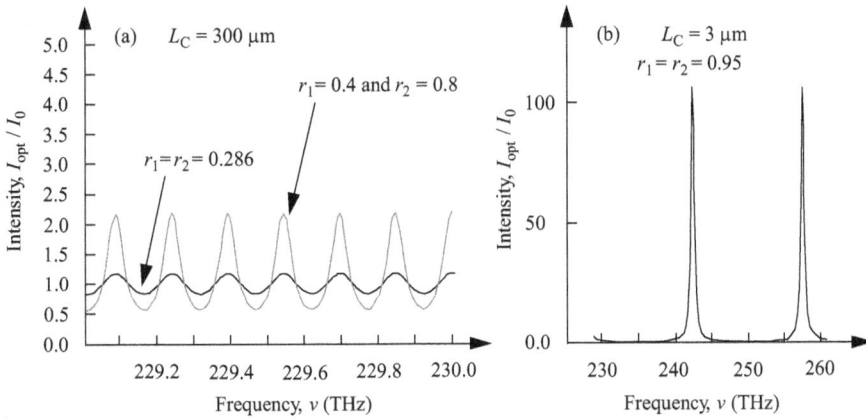

Figure 3.3. (a) Spectral intensity as a function of frequency when $r_1 = r_2 = 0.286$ for photons inside a Fabry–Perot resonant cavity of length $L_C = 300\,\mu$m and dielectric constant $n_r = 3.3$. Optical resonances are spaced by $\Delta\nu = 151$ GHz. The vertical axis is normalized in such a way that $I_0 = 1$ in the calculation. Also shown as the lower curve is the case in which $r_1 = 0.4$ and $r_2 = 0.8$. In this situation, finesse $\mathcal{F} = 2.613$ and $\gamma_{opt} = 58$ GHz. At an optical frequency of $\nu_0 = 229$ THz, this gives an optical of $Q = \nu_0/\gamma_{opt} = 3963$. (b) Spectral intensity as a function of frequency when $r_1 = r_2 = 0.95$ for photons inside a Fabry–Perot resonant cavity of length $L_C = 3\,\mu$m and dielectric constant $n_r = 3.3$. Optical resonances are spaced by $\Delta\nu = 15.1$ THz. In this case, the finesse $\mathcal{F} = \Delta\nu/\gamma_{opt} = 30.6$ and $\gamma_{opt} = 495$ GHz. At an optical frequency of $\nu_0 = 242$ THz, this gives an optical $Q = \nu_0/\gamma_{opt} = 489$.

shown is when $r_1 = 0.4$ and $r_2 = 0.8$, which has a slightly improved finesse of $\mathcal{F} = 2.613$. In this case, $Q = \nu_0/\gamma_{opt} = 3963$ and $\gamma_{opt} = 58$ GHz.

Figure 3.3(b) shows the spectral intensity of a Fabry–Perot optical cavity of length $L_C = 3\,\mu m$, effective refractive index $n_r = 3.3$, mirror reflectivity $r_1 = r_2 = 0.95$, and emission wavelength near $\lambda = 1240$ nm corresponding to a frequency $\nu_0 = 242$ THz. The optical resonator has resonance spacing $\Delta\nu = 15.1$ THz or $\Delta\lambda = 86.7$ nm. Finesse is $\mathcal{F} = \Delta\nu/\gamma_{opt} = 30.6$, linewidth is $\gamma_{opt} = 495$ GHz, and optical $Q = \nu_0/\gamma_{opt} = 489$.

3.1.3 Mode profile in an index-guided slab waveguide

An index-guided slab waveguide structure is created when the refractive index of the active region, n_a, is greater than the refractive index, n_c, of the surrounding material. This, usually small, difference in refractive index acts to guide light close to the active region. Confining light to the active region is important, because only light that overlaps with the active region experiences optical gain and is amplified. The fraction of a Fabry–Perot longitudinal optical resonance in the z-direction that overlaps with the active region is Γ_{opt}. This fraction may be calculated if the optical mode profile in the x- and y-directions of the Fabry–Perot laser diode cavity is known.

The time-independent electromagnetic wave equation, which may be derived from Maxwell's equations assuming no free charge and an electromagnetic wave traveling in the z-direction, can be used to find the mode profile in the passive cavity. In this case

$$\nabla^2 \mathbf{E} = \varepsilon(x, y)k_0^2 \mathbf{E} = 0, \tag{3.20}$$

where $k_0 = \omega/c$ is the propagation constant in free space and $\varepsilon(x, y)$ is the spatially varying dielectric permittivity function in the slab waveguide geometry. This has solution

$$\mathbf{E} = \mathbf{e}\phi\,(y)\psi(x)e^{ikz}, \tag{3.21}$$

where $\mathbf{e}$ is the electric field unit vector and k is the propagation constant in the dielectric.

A solution for the optical confinement factor

$$\Gamma_{opt} = \frac{\int_{y_1}^{y_2} \phi^2(y)\mathrm{d}y}{\int_{-\infty}^{\infty} \phi^2(y)\mathrm{d}y} \tag{3.22}$$

is found by assuming that $\varepsilon(x, y)$ varies slowly in the x-direction compared with the y-direction and by adopting the effective index approximation. In essence, a solution is found for the simple slab waveguide geometry depicted in figure 3.4, in which the thickness of the active region is $t_a = y_2 - y_1$.

Figure 3.5 shows the results of calculating the optical confinement factor of transverse electric (TE) and transverse magnetic (TM) electromagnetic modes in a

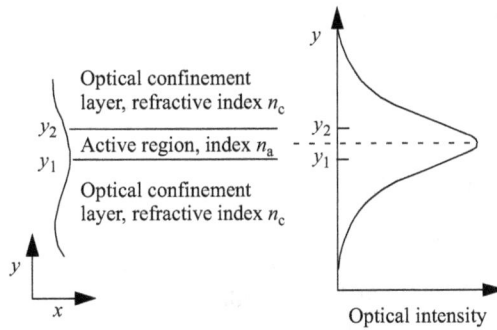

Figure 3.4. Slab waveguide geometry showing the active region of thickness $t_a = y_2 - y_1$ and optical confinement layers. The propagation of light is in the z-direction, which is into the page. Optical intensity peaks in the active region, which has refractive index n_a. The refractive index of the optical confinement layer is n_c.

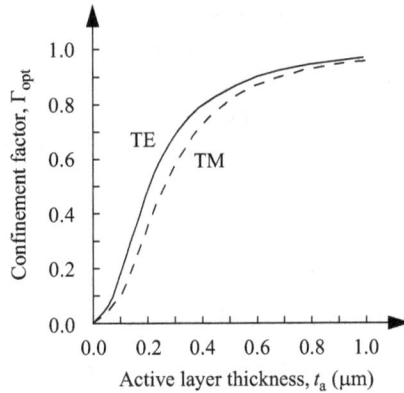

Figure 3.5. Calculated optical confinement factor Γ_{opt} for TE and TM modes in a slab waveguide as a function of bulk-active-layer thickness, t_a. The lasing wavelength is $\lambda_0 = 1310$ nm, the InGaAsP active layer has a refractive index $n_a = n_{r, InGaAsP} = 3.51$, and the InP optical confinement layers have a refractive index $n_c = n_{r, InP} = 3.22$.

slab waveguide as a function of bulk-active-layer thickness. The parameters used are typical for a laser diode with emission at $\lambda_0 = 1310$ nm wavelength and an InGaAsP active region. For a given active-region thickness, the optical confinement factor for TE polarization is greater than that for TM polarization. TE-polarized light propagating in the z-direction has an electric field parallel to the x-direction, and so it is in the plane of the active-region layer.

For most Fabry–Perot laser designs that use index guiding, the ratio of active-region thickness t_a to emission wavelength λ_0 is small, and the confinement factor for TE-polarized light may be found using the approximation [13]

$$\Gamma_{opt, TE} = 2\left(n_a^2 - n_c^2\right)\left(\frac{\pi t_a}{\lambda_0}\right)^2. \tag{3.23}$$

In the case of TM-polarized light, the approximation for the confinement factor is

$$\Gamma_{\text{opt, TE}} = 2\left(n_a^2 - n_c^2\right)\left(\frac{\pi n_c t_a}{n_a \lambda_0}\right)^2. \tag{3.24}$$

3.1.4 Mirror loss and photon lifetime

Previously, in section 3.1.2, reflectivity $r_1 = r_2 = 1$ was used to calculate longitudinal mode frequency in a Fabry–Perot resonator. For most practical situations, it is necessary to consider $r_1 \neq r_2 < 1$. In a typical semiconductor laser diode, the mirror facets have dielectric coatings to precisely control optical power reflectivity values r_1 and r_2.

Rate equation analysis shows that photon *density*, S, grows exponentially in the presence of optical gain $g_{\text{opt}} = -\alpha_{\text{opt}}$, so that $S = S_0 e^{-2\alpha_{\text{opt}} z}$ where the factor of 2 occurs because S is proportional to a field intensity. For steady-state conditions, the rate of growth in S is balanced by the photon loss rate, $1/\tau_{\text{ph}}$, into regions other than the active laser region.

The rate of increase in optical intensity is $G = 2g_{\text{opt}} c/n_r$, where n_r is the effective refractive index and c is the velocity of light in vacuum. For steady-state emission, the photons reflected back to the start location in a single round-trip time $\tau_{\text{RT}} = 2Lcn_r/c$ must have the same photon density. So, if a density of S_0 photons start out from mirror r_1 of the cavity, then $r_2 S_0 e^{(G\tau_{\text{RT}})/2}$ are reflected back from mirror r_2 to grow with another pass down the laser, and $r_1 r_2 S_0 e^{G\tau_{\text{RT}}}$ is reflected from mirror r_1. Hence, in steady-state,

$$S_0 = r_1 r_2 S_0 e^{G\tau_{\text{RT}}}, \tag{3.25}$$

which after taking the logarithm of both sides is

$$G = \frac{1}{\tau_{\text{RT}}} \ln\left(\frac{1}{r_1 r_2}\right) = \frac{c}{2Lcn_r} \ln\left(\frac{1}{r_1 r_2}\right). \tag{3.26}$$

Ignoring spontaneous emission, the rate equation for the photon density is

$$\frac{dS}{dt} = \left(G - \frac{1}{\tau_{\text{ph}}}\right)S = 0 \tag{3.27}$$

giving $G\tau_{\text{ph}} = 1$, so that

$$\frac{1}{\tau_{\text{ph}}} = \frac{c}{2Lcn_r} \ln\left(\frac{1}{r_1 r_2}\right). \tag{3.28}$$

It is necessary to introduce an extra photon loss term to include the possibility of additional absorption and elastic scattering of light. This is done by adding an internal loss rate $1/\tau_i$, so that the total photon loss rate measured in units of s^{-1} is

$$\frac{1}{\tau_{\text{ph}}} = \frac{1}{\tau_i} + \frac{1}{\tau_m} \tag{3.29}$$

or, equivalently,

$$\kappa = \alpha_i + \alpha_m, \tag{3.30}$$

where α_i is the internal photon loss rate, $\alpha_m = 1/\tau_m$ is the mirror photon loss rate, and κ is the total photon loss rate. Often α_i, α_m, and κ are given in units of cm^{-1}, which should be converted to a rate measured in units of s^{-1} for use in the rate equations.

3.1.5 The buried heterostructure Fabry–Perot laser diode

Figure 3.6 is a sketch of a semiconductor, buried heterostructure, Fabry–Perot laser diode. The diagram shows the bulk-active or quantum-well region that runs along the full cavity length and is exposed at one of the two cleaved mirror faces. Carriers are injected into the active region between the two mirrors from the n-type substrate and the p-type epitaxially grown layers that form the p–n junction. Electrical contact with the diode is achieved by depositing a metal film and subsequently alloying it into a surface layer of the semiconductor.

Typically, a bulk-active InGaAsP region has a composition such that its band gap occurs at wavelength $\lambda_g = 1280$ nm. Under lasing conditions, various physical effects cause the lasing wavelength to increase to a longer wavelength, so that the device lases at $\lambda = 1310$ nm. In a typical device, the bulk or multiple-quantum-well active region is $t_a = 0.12$ μm thick, $w_a = 0.8$ μm wide, and has a cavity length L_C created between the two mirror faces. The wafer is thinned before cleaving to form the two mirror facets. Thinning the wafer to about 120 μm thickness helps to ensure that stress-induced irregularities are avoided on the cleaved mirror faces. The buried heterostructure is achieved using an etching and semiconductor regrowth process. Index guiding of the $\lambda = 1310$ nm lasing mode occurs because of the refractive index

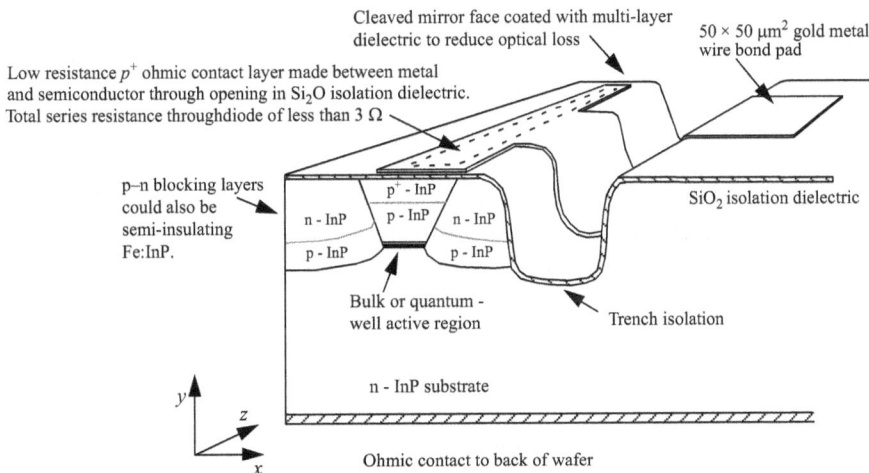

Figure 3.6. Sketch of a typical semiconductor buried heterostructure Fabry–Perot laser diode. The difference in refractive index between the active region and the surrounding dielectric causes index guiding of light propagating in the z-direction.

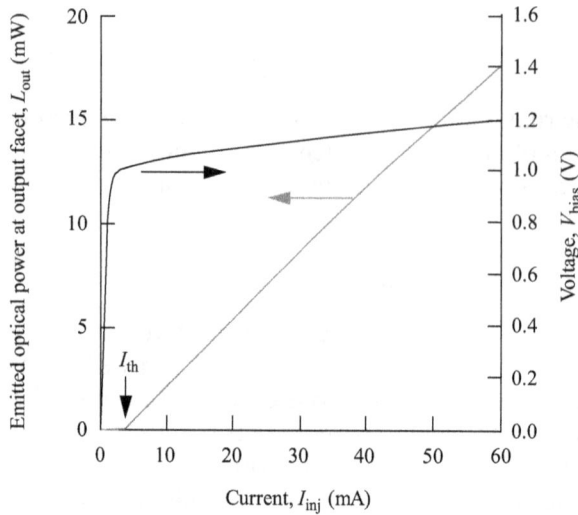

Figure 3.7. LIV measurements of a typical unpackaged InGaAsP/InP buried heterostructure Fabry–Perot laser diode with emission at 1310 nm wavelength. Laser threshold current at room temperature is $I_{th} = 3$ mA and the *single-facet* emitted optical power efficiency is 0.3 mW mA^{-1}. The laser diode has a series resistance of 3 Ω.

difference between the InGaAsP active layer with $n_{r,\ InGaAsP} = 3.51$ and the InP optical confinement layers with refractive index $n_{r,\ InP} = 3.22$. For an index-guided buried heterostructure, this ensures that a single transverse mode and a high optical confinement of greater than $\Gamma_{opt} = 0.25$. A typical Fabry–Perot cavity length is $L_C = 300\ \mu$m. A multi-layer dielectric mirror coating is often used to increase reflectivity to 0.4 on one mirror and 0.8 on the other. This reduces optical loss and reduces laser threshold current to a value that is typically around 3 mA.

Figure 3.7 shows optical output power, L_{out} (red curve), and voltage bias, V_{bias} (black curve), as a function of injection current, I_{inj}, for a typical InGaAsP/InP buried heterostructure Fabry–Perot laser diode at room temperature (20 °C) with emission at 1310 nm wavelength. The LIV characteristics indicate a laser threshold current of $I_{th} = 3$ mA and a single-facet emitted optical power efficiency of 0.3 mW mA^{-1} when $I_{inj} > I_{th}$. The laser diode has a series resistance of 3 Ω. The laser threshold current at a substrate temperature of 85 °C increases to $I_{th} = 13$ mA and *single-facet* emitted optical power efficiency decreases to 0.2 mW mA^{-1}.

A portion of this chapter was reproduced with permission from [14].

Bibliography

[1] Hall R N, Fenner G E, Kingsley J D, Soltys T J and Carlson R O 1962 *Phys. Rev. Lett.* **9** 366
[2] Hayashi I, Panish M B, Foy P W and Sumski S 1970 *Appl. Phys. Lett.* **17** 109
[3] Soda H, Iga K, Kitahara C and Suematsu Y 1977 *Jpn. J. Appl. Phys.* **18** 2329
[4] For a review, see Kumagai O, Ikeda M and Yamamoto M 2013 *Proc. IEEE* **101** 2243
[5] Iga K, Oikawa M, Misawa S, Banno J and Kokubun Y 1982 *Appl. Opt.* **21** 3456
[6] McCall S L, Levi A F J, Slusher R E, Pearton S J and Logan R A 1992 *Appl. Phys. Lett.* **60** 289

[7] Kogelnik H and Shank C V 1971 *Appl. Phys. Lett.* **18** 152
     Kogelnik H and Shank C V 1973 *J. Appl. Phys.* **43** 2327
[8] Morton P A 1994 *Appl. Phys. Lett.* **64** 2634
[9] Stern B, Ji X, Okawachi Y, Gaeta A L and Lipson M 2018 *Nature* **562** 401
[10] Hsu C W, Zhen B, Stone A D, Joannopoulos J D and Soljačić M 2016 *Nat. Rev. Mater.* **1** 16048
[11] Asbóth J K, Oroszlány L and Pályi A 2016 *A Short Course on Topological Insulators: Lecture Notes in Physics* **919** (Cham: Springer)
[12] For example Saleh B E A and Teich M C 1991 *Fundamentals of Photonics* (New York: Wiley)
[13] Dumpke W P 1975 *IEEE J. Quantum Electron.* **11** 400
[14] Levi A F J 2023 *Applied Quantum Mechanics* 3rd edn (Cambridge: Cambridge University Press)

IOP Publishing

Essential Semiconductor Laser Device Physics (Second Edition)

A F J Levi

Chapter 4

Single-mode rate equations

This chapter introduces the continuum mean-field single-mode semiconductor laser diode rate equations. Parameterization of the carrier-recombination rate, carrier-dependent optical gain and optical gain saturation, carrier-dependent spontaneous emission, and the fraction of spontaneous emission feeding the single laser mode are discussed. Calculation of the steady-state and large-signal transient response using fourth-order Runge–Kutta numerical integration is demonstrated. Laser threshold, relaxation oscillations, and turn-on delay are described, as well as the generation of optical pulses by gain switching, Q-switching, and mode-locking. The essential elements of scaling laser diode size by reducing the fraction of spontaneous emission feeding the single laser mode, the concept of critical slowing associated with a nonequilibrium phase transition, cavity formation, and small-signal intensity response are also covered.

To understand the operation of a semiconductor buried-heterostructure Fabry–Perot laser diode, it is helpful to develop a model. Any such model should consider the fact that electrons are driven into the device as injection current and photons emitted into the lasing mode contribute to an electromagnetic field. The electrons and photons form a bipartite nonequilibrium system whose detailed description of behavior can be quite complicated. A simplified approach that captures many aspects of device operation uses continuum mean-field single-mode rate equations. Such a rate equation model is useful to assist understanding and intuition about semiconductor laser diode behavior, including the large-signal transient response.

4.1 Continuum mean-field single-mode semiconductor laser diode rate equations

For ease of calculation, it is assumed that lasing occurs into only one optical mode of radial frequency ω_s. This is the *single-mode* approximation. Further simplification is achieved by not incorporating any variation in optical gain, optical loss, or carrier density along the longitudinal (z) axis. This is equivalent to a lumped-element model

doi:10.1088/978-0-7503-6417-1ch4
4-1

in which only the average field, or *mean field*, is used. Typically, this rate equation model adopts approximations that parameterize optical gain, spontaneous emission, and nonradiative recombination. The conduction band and valence band electrons are assumed to be non-interacting, have the same density, and are thermalized so that they may be characterized by a single temperature. The electron and photon *density* (rather than quantized particle number) is calculated because it is convenient to integrate a continuous and smooth function.

When considering the rate equations, care should be taken to define the parameters used. The current, I_{inj}, injected into a typical diode illustrated in figure 4.1 is measured in amps, and the volume of the active region into which electrons flow is V_{vol}. The carrier density in the active region is n and is measured in units of either m^{-3} or cm^{-3}. The photon density in the optical mode of frequency ω_s is S and is measured in units of m^{-3} or cm^{-3}. The two mirrors used to form the optical cavity for photons in the device have reflectivity r_1 and r_2, respectively. The optical loss at frequency ω_s in the cavity is κ and is measured in m^{-1} or cm^{-1}. The continuum mean-field rate equations require that the fraction of spontaneous emission, β, feeding into the lasing mode at frequency ω_s is known. It turns out that this is somewhat difficult to define in an active device and β is best considered a phenomenological parameter. In a large Fabry–Perot device with cavity length $L_C = 300 \ \mu m$, a typical value for β is in the range $10^{-4} < \beta < 10^{-5}$. In devices with smaller L_C, the value of β becomes larger. Theoretically, the maximum possible value of β is unity.

Rate equations are used to describe the flow of particle density in and out of a region of interest. The physical quantities monitored are carrier density n, photon density S in a single optical mode of frequency ω_s, and current I_{inj} as a function of time, t. Rate equations keep track of current, carrier, and photon flow in and out of the device as it is driven away from equilibrium operating conditions.

A *bucket model* may be used to represent the active region of the semiconductor (see figure 4.2). One may imagine charge carriers supplied by a current I_{inj} being

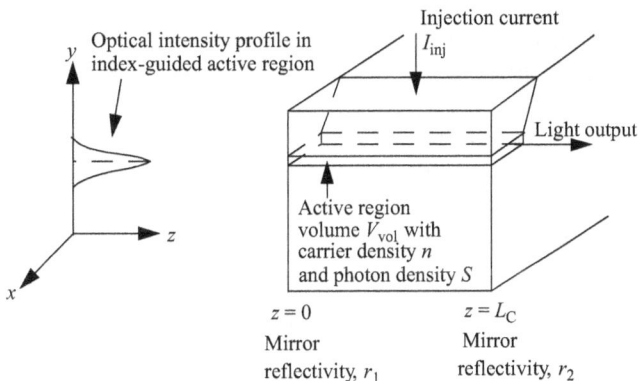

Figure 4.1. Section through a buried-heterostructure Fabry–Perot laser diode. Index guiding ensures that the optical intensity is tightly confined near the active region of the device. The Fabry–Perot cavity length is L_C and the mirror reflectivity is r_1 at $z = 0$ and r_2 at $z = L_C$. When current I_{inj} is injected into the diode, the active region of volume V_{vol} has carrier density n and photon density S.

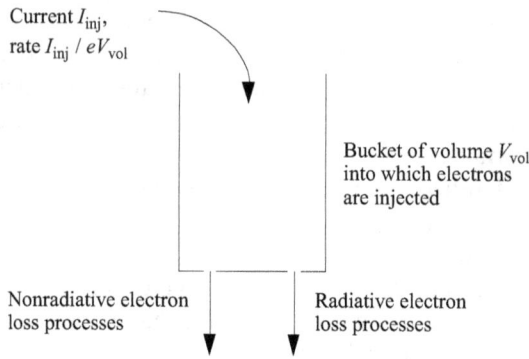

Figure 4.2. Illustration of the bucket model for electron rates into and out of a semiconductor active region.

poured into the bucket at a rate so that the *density* of electrons per second increases as I_{inj}/eV_{vol}, where e is the electron charge and V_{vol} is the volume of the active region. There are losses or leaks in the bucket, which represent mechanisms for removing electrons from the system. Electrons can be removed by emitting a photon or by nonradiative processes.

In its simplest form, two first-order coupled rate equations are considered.

One equation describes the rate of change in carrier density dn/dt in the device. Carrier density will increase as more current is injected, so dn/dt is expected to have a term proportional to the current I_{inj}. Carriers may be removed from the active region of the device by nonradiative and spontaneous photon-emission carrier-recombination processes. This carrier loss is characterized by the carrier lifetime τ_n, so that dn/dt is proportional to $-n/\tau_n$. The negative sign reflects the fact that carriers are removed from the system. There are also carrier losses due to stimulated photon emission, in which an electron in the conduction band and a hole in the valence band are removed to create a photon in the lasing mode. This stimulated emission process influences the number of carriers via the rate $-GS$, where G is the optical gain and S is the photon density in the lasing mode.

A second equation describes the rate of change in photon density dS/dt in the device. Photon density increases due to the presence of optical gain, so the term is proportional to GS. There are also optical losses that a total optical loss rate can describe $\kappa = 1/\tau_{ph}$, giving a term $-\kappa S$. Finally, there is a fraction β of total spontaneous emission r_{spon} feeding into the lasing mode that makes a contribution βr_{spon}.

These considerations result in two coupled rate equations that describe the behavior of the laser diode:

$$\frac{dn}{dt} = \frac{I_{inj}}{eV_{vol}} - \frac{n}{\tau_n} - GS \tag{4.1}$$

$$\frac{dS}{dt} = (G - \kappa)S + \beta r_{spon}. \tag{4.2}$$

Equations (4.1) and (4.2) are the continuum mean-field single-mode rate equations. In equation (4.1), $1/\tau_n$ is a phenomenological carrier-recombination rate, where

$$\frac{1}{\tau_n} = A_{nr} + Bn + Cn^2. \tag{4.3}$$

Here, A_{nr} is the nonradiative recombination rate, B is the spontaneous emission rate, and C is a higher-order term. The total spontaneous emission rate in the device is taken to be

$$r_{spon} = Bn^2 \tag{4.4}$$

and the function for optical gain in a bulk-active region device is approximated as

$$G_{bulk} = \Gamma_{opt} \frac{G_{slope}(n - n_{ot})}{1 + \varepsilon_{bulk} S}, \tag{4.5}$$

where G_{slope} is the differential optical gain with respect to carrier density, Γ_{opt} is the optical confinement factor, and ε_{bulk} is the gain saturation coefficient in a bulk semiconductor. For $\varepsilon_{bulk} S \ll 1$, equation (4.5) may be written as $G_{bulk} = \Gamma_{opt} G_{slope}(n - n_{ot})(1 - \varepsilon_{bulk} S)$.

For a quantum-well active region, the optical gain function may be approximated by

$$G_{QW} = \Gamma_{opt} \frac{G_{const}}{(1 + \varepsilon_{QW} S)} \left(\ln \left(\frac{n}{n_{ot}} \right) \right), \tag{4.6}$$

where G_{const} is a constant determined from experiment. The carrier density needed to achieve optical transparency at radial frequency ω_s is n_{ot}. In the absence of any additional sources of noise, the spectral linewidth of a high finesse Fabry–Perot resonator at transparency is $\gamma_{opt} = \Delta \nu / \mathcal{F}$ and $Q = \nu_0 / \gamma_{opt}$. The gain saturation terms ε_{bulk} and ε_{QW} become important at high optical intensities in the cavity. In practice, the gain functions G_{bulk} and G_{QW} are quite accurate descriptions of the dependence of peak gain on carrier density.

It is now possible to explain the steady-state carrier density and photon density characteristics of a laser diode as a function of injected current. Equation (4.1) for the steady-state case is

$$\frac{dn}{dt} = \frac{I_{inj}}{e V_{vol}} - \frac{n}{\tau_n} - GS = 0. \tag{4.7}$$

This equation shows that in a steady-state, the rate of electron density injected into the active region is exactly balanced by the removal of electrons via the recombination rate $Q = n/\tau_n$ and the optically stimulated recombination rate GS.

In a steady-state, equation (4.2) is

$$\frac{dS}{dt} = (G - \kappa)S + \beta r_{spon} = 0, \tag{4.8}$$

which may be re-written as

$$S = \frac{\beta r_{\text{spon}}}{(\kappa - G)}. \tag{4.9}$$

It is this last equation that may be used as a starting point to explain how a laser works. The numerator βr_{spon} on the right-hand side of equation (4.9) shows that the optical output S of the laser amplifier is fed by a small fraction of the total spontaneous emission in the device. Because spontaneous emission is a stochastic (random) quantum-mechanical process, the laser may be viewed as amplifying noise. This source of noise has an impact on the performance of laser diodes and is of particular interest in applications such as high-performance fiber-optic communication systems.

The denominator $(\kappa - G)$ on the right-hand side of equation (4.9) is the term responsible for optical amplification and lasing emission. As electrons are injected into the device, optical gain increases, and $(\kappa - G)$ approaches zero, the amplification of spontaneous emission increases. This increase in photon density in the device is so great that the stimulated carrier-recombination rate term $-GS$ in equation (4.7) becomes large. As $(\kappa - G)$ continues to approach zero, the net optical amplification of spontaneous emission $1/(\kappa - G)$ becomes very large, and the stimulated recombination rate $-GS$ dominates equation (4.7). Under these operating conditions, essentially every additional electron injected by the current I_{inj} into the active region recombines very rapidly to create an additional photon in the lasing mode at radial frequency ω_s.

Figure 4.3 illustrates this situation. As optical gain approaches a minimum in Fabry–Perot optical-cavity loss such that $(G - \kappa) \rightarrow 0$, the lasing mode will coincide with that of the Fabry–Perot resonance of frequency ω_s nearest to the peak optical gain. Because, in this case, only one high-Q optical-cavity resonance is in the same frequency range as semiconductor optical gain, lasing occurs in one optical mode at a radial frequency ω_s.

As stimulated recombination begins to dominate, light emission at radial frequency ω_s increases rapidly. The point at which this occurs is called the laser threshold. Associated with the laser diode threshold is a threshold current, I_{th}, and a threshold carrier density, n_{th}. For currents above the laser threshold, carrier density does not increase very much because the stimulated recombination rate $-GS$ dominates the carrier dynamics described by equation (4.7). The carrier density is said to be *pinned* above threshold. Such carrier pinning results in a rapid linear increase in laser light output intensity with increasing injection current because almost every extra injected electron is converted to a lasing photon. The large recombination rate $-GS$ that occurs above the laser threshold current, I_{th}, results in a characteristic -3 dB light output small-signal current modulation frequency response that can be as high as a few tens of GHz.

The steady-state characteristics described in the previous few paragraphs are illustrated in figures 4.4 and 4.5. Figure 4.4 shows the total laser diode light output power, L_{out}, as a function of injected current, I_{inj}, for a device with a threshold

Light in lasing mode becomes large as $(G - \kappa) \to 0$.

In steady state, $dS/dt = (G - \kappa)S + \beta r_{spon} = 0$, so that

$$S = \beta r_{spon}/(\kappa - G)$$

Optical loss, κ is a function of frequency in Fabry-Perot resonator

Optical gain when $I_{inj} = I_{th}$

Optical transparency for bulk active region when lasing

Optical gain, $G(\omega)$

0

Optical frequency, ω

Lasing mode frequency

Equally spaced resonant modes of the Fabry-Perot cavity

Optical gain when $I_{inj} \ll I_{th}$

Optical gain when $I_{inj} < I_{th}$

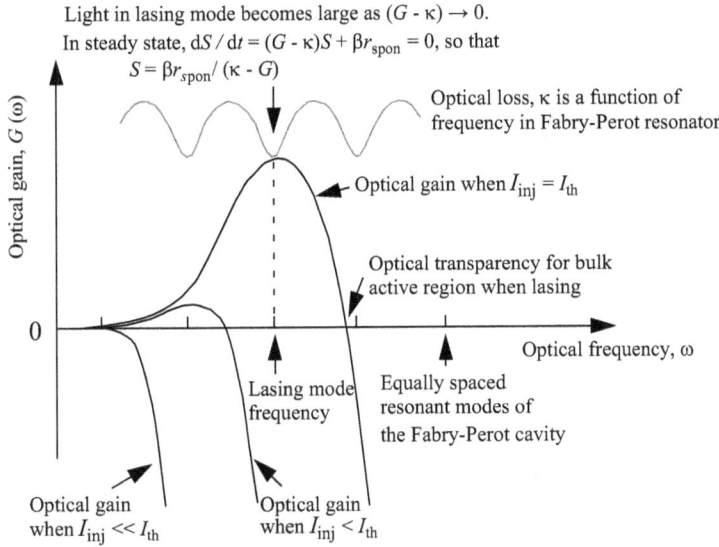

Figure 4.3. Schematic plot of optical gain as a function of optical frequency. Lasing will occur when optical gain approaches optical loss. This will usually occur at a resonance of the Fabry–Perot resonator, the optical loss of which is also shown.

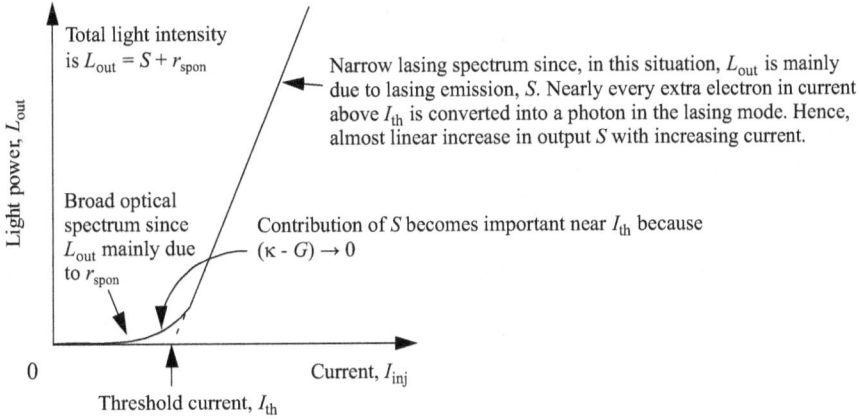

Total light intensity is $L_{out} = S + r_{spon}$

Narrow lasing spectrum since, in this situation, L_{out} is mainly due to lasing emission, S. Nearly every extra electron in current above I_{th} is converted into a photon in the lasing mode. Hence, almost linear increase in output S with increasing current.

Light power, L_{out}

Broad optical spectrum since L_{out} mainly due to r_{spon}

Contribution of S becomes important near I_{th} because $(\kappa - G) \to 0$

0

Current, I_{inj}

Threshold current, I_{th}

Figure 4.4. Anticipated total light output power, L_{out}, as a function of injected current, I_{inj}. The device has a threshold current I_{th}.

current I_{th}. Figure 4.5 shows carrier density n as a function of injected current I_{inj}. As may be seen, carrier density is pinned to a value of approximately n_{th} when the current is greater in value than I_{th}.

The single-mode light output power from the mirror of a Fabry–Perot laser diode with reflectivity r_1 may be calculated using

$$L_{out} = \hbar \omega_s \alpha_m \frac{c}{n_r} V_{vol} \left(\frac{1 - r_1}{2 - r_1 - r_2} \right) S. \tag{4.10}$$

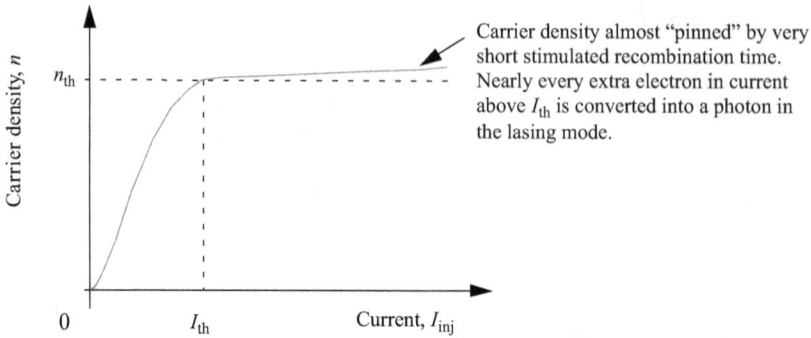

Figure 4.5. Anticipated carrier density n as a function of injected current I_{inj}. The carrier density is pinned to a value of approximately n_{th} when the current is greater than the threshold current, I_{th}.

4.2 Numerical method for solving rate equations

In the previous section, the single-mode laser rate equations were used to qualitatively predict the steady-state behavior of a semiconductor laser diode. While many practical applications can make use of such steady-state characteristics, there is also interest in the high-speed, small-signal, and large-signal response of the device. For example, conventional data transmission via an optical fiber medium requires a photon source in which lasing light intensity is modulated. A conceptually straightforward way to control lasing light emission is to change the injection current. Modulating in a one-bit digital fashion, a high light level might correspond to binary 1, and a low light level might correspond to binary 0. To efficiently pass data in a fiber-optic link, it is important to know how fast the light output of a laser can switch from a 1 state to a 0 state and vice-versa. To answer such basic questions concerning laser performance, it is often helpful to resort to numerical methods that are capable of predicting the large-signal dynamic response of a laser diode's carrier density and laser light emission in response to rapid changes in injection current. The numerical solution to the coupled continuum mean-field rate equations (4.1) and (4.2) requires an accurate integrator such as the fourth-order Runge–Kutta method described in the next section.

4.2.1 The Runge–Kutta method

Ordinary differential equations may always be expressed in terms of first-order differential equations. For example,

$$\frac{d^2 y}{dt^2} + a(t)\frac{dy}{dt} = b(t) \tag{4.11}$$

may be written as two first-order coupled differential equations

$$\frac{dy}{dt} = f(t) \tag{4.12}$$

$$\frac{df(t)}{dt} = b(t) - a(t)f(t). \tag{4.13}$$

It follows that one is interested in the study of N-coupled first-order differential equations for the functions y_j having the form

$$\frac{d}{dt}y_j(t) = f_j(t, y_1, y_2, \ldots, y_N), \tag{4.14}$$

where the right-hand side is a known function and $j = 1, 2, \ldots, N$.

After applying boundary conditions and an initial value, a finite step t_{step} may be used to solve the equations numerically. For example, Euler's method

$$y_{n+1} = y_n + t_{step}f(t_n, y_n) + O\left(t_{step}^2\right) \tag{4.15}$$

might be used to advance the solution from t_n to $t_{n+1} \equiv t_n + t_{step}$. It evolves the solution through an interval t_{step}, using only derivative information contained in $f(t,y)$ at the beginning of the interval. Unfortunately, the error in each step is $O(t_{step}^2)$, which is only one power of t_{step} smaller than the estimate function $t_{step}f(t_n, y_n)$.

It is possible to do better than this by first taking a trial step to the midpoint of the interval and then using the value of both t and y at the midpoint to compute a more accurate step across the entire interval. If $f(t, y)$ is evaluated in such a way that first-order and some higher-order terms cancel, it is possible to make a very accurate numerical integrator. The fourth-order Runge–Kutta method does just this [1–3]. As illustrated in figure 4.6, each step along $f(t, y)$ is evaluated four times: once at the initial point, twice at the mid-points, and once at the trial point.

The fourth-order Runge–Kutta method is summarized by the equation

$$y_{n+1} = y_n + \frac{k_1}{6} + \frac{k_2}{3} + \frac{k_3}{3} + \frac{k_4}{6} + O\left(t_{step}^5\right), \tag{4.16}$$

where

$$k_1 = t_{step}f(t_n, y_n) \tag{4.17}$$

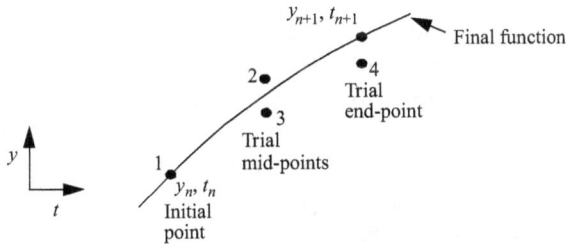

Figure 4.6. Illustration of the fourth-order Runge–Kutta method of numerically estimating integration of a function. The solution is advanced through the time interval $t_{step} = t_{n+1} - t_n$.

is used to evaluate at the initial point,

$$k_2 = t_{step} f\left(t_n + \frac{t_{step}}{2}, y_n + \frac{k_1}{2}\right) \tag{4.18}$$

estimates the midpoint using $k_1/2$,

$$k_3 = t_{step} f\left(t_n + \frac{t_{step}}{2}, y_n + \frac{k_2}{2}\right) \tag{4.19}$$

estimates the midpoint using $k_2/2$, and

$$k_4 = t_{step} f\left(t_n + t_{step}, y_n + k_3\right) \tag{4.20}$$

evaluates the endpoint using k_3.

The single-mode rate equations (4.1) and (4.2) are two coupled first-order differential equations that may be solved using the fourth-order Runge–Kutta method. When doing so, care should be taken to make sure that the estimates for trial points k_1, k_2, k_3, and k_4 are updated using the most current values.

4.3 Large-signal transient response

When writing a computer program to model the behavior of a semiconductor laser diode with a bulk-active gain region, it is necessary to define the functions in the coupled rate equations (4.1) and (4.2) and assign numerical values to parameters. Table 4.1 gives typical parameter values for laser diodes with photon emission at the indicated wavelengths. The stiffness of the coupled rate equations to be numerically integrated depends on the parameter values, and this impacts the choice of time step values. The stiffer the equations, the shorter the time step. In practice, a constant time step value of $t_{step} = 1$ ps works well. However, more sophisticated routines with adaptive step sizes may also be implemented.

The parameter values in table 4.1 give rise to the indicated laser diode threshold current values, I_{th}. Direct band gap semiconductors such as GaAs have a larger band gap energy than InGaAsP and so have lasing emission at shorter wavelengths. They also have a larger optical gain-slope coefficient, G_{slope}, and typically a lower value of laser threshold current. In table 4.1, the laser diode with photon emission at a 850 nm wavelength has a laser threshold current about one-quarter of the value for the device with emission at the 1550 nm wavelength.

A measure of the efficiency by which current injected into a laser diode is converted to lasing photon emission is the quantity η_{slope}. This slope efficiency is the change in *total* emitted lasing optical power with increasing injection current measured from the laser threshold current. For a typical semiconductor laser diode, slope efficiency is measured in units of mW emitted lasing optical power per mA of additional injection current. The larger optical gain-slope coefficient, G_{slope}, and relatively smaller nonradiative and optical losses in GaAs compared to InGaAsP result in GaAs laser diodes that typically have a larger slope efficiency and are more effective at converting current to lasing light emission. In table 4.1, the slope

Table 4.1. Fabry–Perot laser diode rate equation parameters.

Description	Parameter	GaAs 850 nm wavelength	InGaAsP 1310 nm wavelength	InGaAsP 1550 nm wavelength
Refractive index	n_r	3.3	4	4
Cavity length	L_C (cm)	250×10^{-4}	300×10^{-4}	500×10^{-4}
Active layer thickness	t_a (cm)	0.14×10^{-4}	0.14×10^{-4}	0.14×10^{-4}
Active layer width	w_a (cm)	0.8×10^{-4}	0.8×10^{-4}	0.8×10^{-4}
Integration time increment	t_{inc} (s)	1×10^{-12}	1×10^{-12}	1×10^{-12}
Nonradiative recombination coefficient	A_{nr} (s^{-1})	2×10^{8}	2×10^{8}	1×10^{8}
Radiative recombination coefficient	B ($cm^3\ s^{-1}$)	1×10^{-10}	1×10^{-10}	1×10^{-10}
Nonlinear recombination coefficient	C ($cm^6\ s^{-1}$)	1×10^{-29}	1×10^{-29}	5×10^{-29}
Transparency carrier density	n_{ot} (cm^{-3})	1×10^{18}	1×10^{18}	1×10^{18}
Optical gain-slope coefficient	G_{slope} ($cm^2\ s^{-1}$)	3.3×10^{-16}	2.5×10^{-16}	2.0×10^{-16}
Gain saturation coefficient	ε_{bulk} (cm^3)	2×10^{-18}	3×10^{-18}	5×10^{-18}
Spontaneous emission coefficient	β	1×10^{-4}	5×10^{-5}	1×10^{-5}
Optical confinement factor	Γ_{opt}	0.25	0.25	0.25
Mirror 1 reflectivity	r_1	0.3	0.32	0.32
Mirror 2 reflectivity	r_2	0.3	0.32	0.32
Internal optical loss	α_i (cm^{-1})	20	40	50
Threshold current	I_{th} (mA)	3.4	5.7	14.2
Slope efficiency	η_{slope} (mW mA^{-1})	1.03	0.46	0.25

efficiency of the laser diode with photon emission at the 850 nm wavelength is four times greater than the device with emission at the 1550 nm wavelength.

In figure 4.7, the results of calculating the large-signal response of a diode laser from the zero injection current off-state to a 30 mA step increase in current is shown (current $I_{inj}(t < 0) = 0$ mA and $I_{inj}(t \geqslant 0) = 30$ mA). The model uses $t_{step} = 1$ ps and parameters are given in table 4.1 for a laser with an emission wavelength near $\lambda_0 = 1310$ nm.

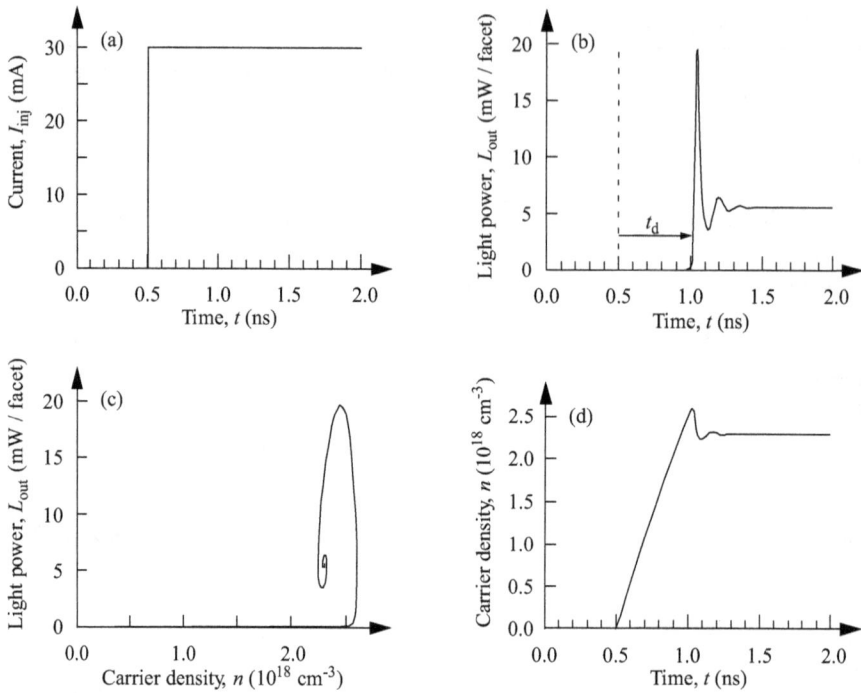

Figure 4.7. (a) Input step current as a function of time. The current is initially zero, increasing abruptly to 30 mA at time $t = 0.5$ ns. The laser threshold is $I_{th} = 5.7$ mA. (b) Laser diode light output power per mirror facet as a function of time, showing turn-on delay, t_d, and overshoot in optical light output. The mirror reflectivity is the same for each facet. (c) Laser diode light output power per mirror facet as a function of carrier density. (d) Conduction band and valence band carrier density as a function of time. The calculation uses parameters for an InGaAsP device given in table 4.1 with emission at the 1310 nm wavelength.

There are a number of characteristic features worth mentioning. First, because the device is turned on from a zero-current (off) state, there is a significant *turn-on delay*, t_d, which is the time to reach half the steady-state optical output level. This is partly because it takes time to inject enough carriers to bring the device to a lasing state. Second, as usual with coupled rate equations, there is a (phase) delay between carrier density, n, and photon density, S. The electrons *lead* the photon density. This gives rise to significant photon density overshoot and *relaxation oscillations* in both the optical output, L_{out}, and the carrier density, n, as the system tries to establish steady-state conditions. This is typical of the temporal response to a large step change in injection current, especially if the current, I_{inj}, passes through the threshold value, I_{th}. Relaxation oscillations limit the useful large-signal digital switching rate of laser diodes.

4.3.1 Optical pulse generation

The large-signal response of a semiconductor laser diode can be used to create short optical pulses. The mechanisms by which this is done include gain switching, Q-switching, and mode-locking [4].

The overshoot in carrier density relative to the steady-state value ($n > n_{\text{th}}$) followed by an overshoot in laser light output relative to the steady-state value ($L_{\text{out}} > L_{\text{out}}(t \rightarrow \infty)$) shown in figures 4.7(d) and (b), respectively, in response to a step change in injection current, can be exploited to create a sequence of high-intensity optical pulses from the laser diode. Simply applying a large amplitude injection current modulated at a radial frequency ω_{mod} to a typical semiconductor laser diode can result in the emission of an optical pulse with a pulse width of a few tens of picoseconds every modulation period, $\tau_{\text{mod}} = 2\pi/\omega_{\text{mod}}$. This is possible because the carrier density overshoot provides a reservoir of excited electronic states in excess of the steady-state value and an optical gain larger than the steady-state value. After the first spontaneous emission event into the laser mode, excited electronic states can be efficiently and rapidly converted into photons to create an optical pulse via the GS stimulated emission term in equation (4.8).

This basic gain-switching approach can be engineered to be more effective by, for example, introducing an intracavity saturable absorber to modulate the optical Q of a laser diode. At low light levels, the absorber introduces loss in the cavity of a laser diode driven by a constant injection current and the carrier density can increase to a high level. Spontaneous emission initiates lasing, and the light level in the cavity becomes so large that the absorber saturates, decreasing optical loss in the cavity and increasing optical Q. This Q-switching mechanism can be passive or active. In the active case, an intracavity absorber is modulated externally. Of relevance for digital control, light output as a function of voltage applied to the absorber exhibits hysteresis [5]. Lasing light emission can be synchronized with a control signal or clock to facilitate system integration. Applications include using intracavity saturable absorbers to perform digital transmission [6] and logic functions [7].

Optical pulses shorter than those achieved by Q-switching can be created in a mode-locked laser diode. In this case, the optical gain bandwidth is designed to be large so that many modes can lase. In a typical passive device with a linear optical-cavity geometry, the injection current is constant, and the relative phase difference between adjacent longitudinal lasing modes is controlled to be fixed in value so that the total pulsed photon field from the comb of lasing modes with constant emission per mode is periodic in time. An actively mode-locked device provides additional control over the lasing frequency comb by modulating optical gain at the frequency difference between cavity modes. The light output pulse width from an active mode-locked laser diode can be sub-picosecond and proportional to the inverse of the optical gain bandwidth. Applications of mode-locked lasers include the generation of microwave signals from frequency combs [8] and many other opportunities for integration into systems [9]. Separately, tunable low-noise microwave and mm-wave clock signals of high spectral purity can be generated by mixing laser diode-pumped comb sources in a detector [10–12].

4.3.2 Scaling with spontaneous emission factor, β

The fraction of spontaneous emission, β, feeding into the laser mode depends on the size of the active region, optical-cavity design details, and laser operating conditions.

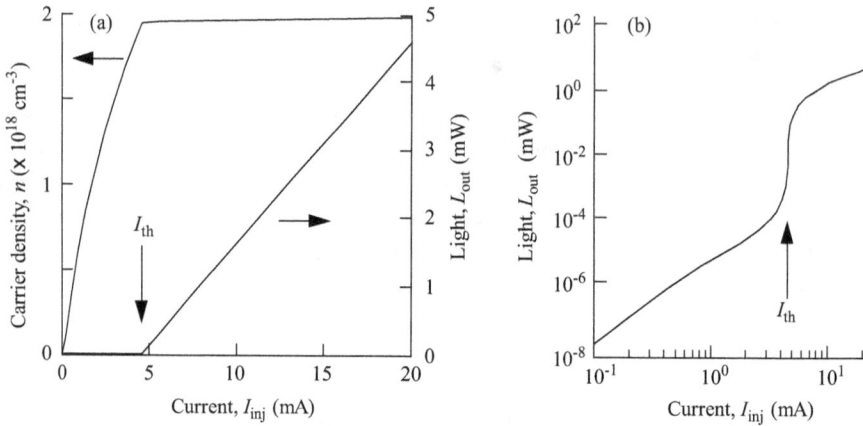

Figure 4.8. (a) Typical Fabry–Perot laser diode light emission and carrier density as a function of injection current, I_{inj}. The carrier density is pinned at about $n_{th} = 2 \times 10^{18}$ cm^{-3} for an injection current above the laser threshold value $I_{th} = 4.6$ mA. (b) A log–log plot of $L_{out} - I_{inj}$ shown in (a). The parameters used are those for an InGaAsP device given in table 4.1 with emission at the 1310 nm wavelength, $r_1 = 0.32$, and $r_2 = 0.98$.

Figure 4.8 shows the results of using continuum mean-field rate equations to calculate light output power, L_{out}, and carrier density, n, as a function of drive current for a typical Fabry–Perot laser diode. In figure 4.8(a), the laser diode threshold current, I_{th}, is determined by linearly extrapolating the high-slope portion of the curve to $L_{out} = 0$. As may be seen, the threshold current is $I_{th} = 4.6$ mA. The $n - I_{inj}$ plot shows that carrier density is pinned for injection current above the threshold and has a value of about $n_{th} = 2 \times 10^{18}$ cm^{-3}. The log–log $L_{out} - I_{inj}$ curve shown in figure 4.8(b) is particularly useful for showing the behavior of the below-threshold light level. For the device parameters used, the laser threshold is associated with a very rapid change in optical output power near $I_{inj} = I_{th}$.

The spontaneous emission factor β can be used to predict behavior as laser diode geometry is scaled to smaller device dimensions. Typically, β will increase with such device scaling.

Assuming the laser diode optical cavity may be modified such that β increases, but *no* other parameters change, figure 4.9 shows the predicted (a) $L_{out} - I_{inj}$ and (b) $n - I_{inj}$ behavior on a log–log plot for $\beta = 10^{-4}$, $\beta = 10^{-3}$, $\beta = 10^{-2}$, $\beta = 10^{-1}$, and $\beta = 1$.

The measure of laser threshold that assumes an abrupt increase in light output with increasing injection current becomes less useful as device geometry is modified to increase the spontaneous emission coefficient, β. In the extreme case, when $\beta = 1$ all spontaneous emission feeds into the laser mode, and evidence of a laser threshold in the $L_{out} - I_{inj}$ characteristic is lost. Likewise, carrier pinning does not occur when $\beta = 1$. If the existence of a laser threshold in a macroscopic configuration is associated with a second-order nonequilibrium phase transition (between disordered light below threshold and ordered light above threshold), then, in any device geometry, as β approaches unity, the transition is no longer evident in either the

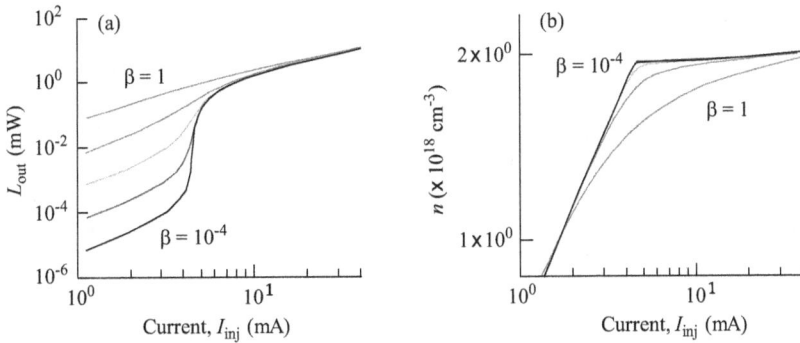

Figure 4.9. (a) Fabry–Perot laser diode light emission and (b) carrier density as a function of injection current, I_{inj}, for values of spontaneous emission factor $\beta = 10^{-4}$, $\beta = 10^{-3}$, $\beta = 10^{-2}$, $\beta = 10^{-1}$, and $\beta = 1$. Note the use of logarithmic scales. The parameters in the calculations are those for an InGaAsP device given in table 4.1 with emission at the 1310 nm wavelength, $r_1 = 0.32$, and $r_2 = 0.98$.

$L_{out} - I_{inj}$ or the $n - I_{inj}$ characteristics. These conclusions are not substantially modified when an increase in β is achieved by scaling (decreasing the size of) the laser diode using all available experimentally accessible device design parameters.

4.3.3 Critical slowing

A time delay t_d exists between the onset of a step change in diode forward current from zero current at time $t = 0$ and laser light output L_{out} to reach half of the steady-state value. The results of calculating the time delay t_d as a function of step current in the range $0 < I_{step} < 20$ mA for a typical macroscopic InGaAs laser diode are shown in figure 4.10(a). There is an increase in delay for values of current near threshold current, $I_{th} = 4.6$ mA. This characteristic behavior is called critical slowing. Figure 4.10(b) is the same as (a) but plotted on a log–log scale.

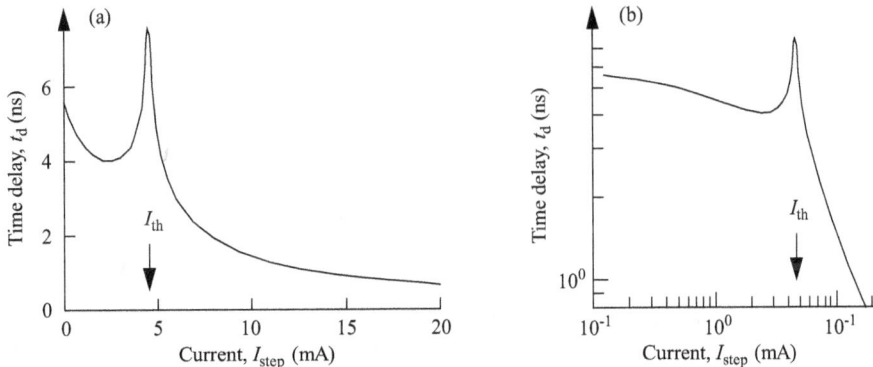

Figure 4.10. (a) Time delay t_d as a function of step current, I_{step}. Threshold current, $I_{th} = 4.6$ mA. (b) Same as (a) but plotted on a $\log_{10}$ scale. The parameters used in calculations are those for the InGaAsP device given in table 4.1 with emission at the 1310 nm wavelength, $r_1 = 0.32$, and $r_2 = 0.98$.

4.3.4 The origin of critical slowing

Critical slowing is associated with the presence of a phase transition. Formally, a phase transition only exists in the large particle number limit (the thermodynamic limit) and so this description can only be applied to a macroscopic system. The laser threshold in a macroscopic device occurs at a second-order nonequilibrium phase transition in which the magnitude of the photon field acts as the order parameter [13] and critical slowing is due to long-lived fluctuations [14]. Random spontaneous emission suppresses long-lived fluctuations and so acts to damp critical slowing. For this reason, increasing the spontaneous emission factor, β, reduces the off–on time delay, t_d, in response to a step increase in injection current from $I_{inj} = 0$ mA to $I_{inj} = I_{step} > I_{th}$.

It is worth noting that a laser diode with a large, near-unity value of β does *not* have a faster above-threshold *small-signal* modulation response compared to a conventional macroscopic device with $\beta \ll 1$. The reason for this may be understood by considering the coupled equations (4.1) and (4.2). With increasing β carriers are less well pinned and this results in the stimulated recombination rate $-GS$ in equation (4.1) becoming less effective in determining the temporal response. Hence, a device with a large value of β can have a modulation response that is closer to that of a light-emitting diode (LED) rather than a conventional macroscopic laser diode.

In figure 4.11(a) the time delay, t_d, associated with critical slowing is shown as a function of step current, I_{step}, for different values of the spontaneous emission factor $\beta = 10^{-4}$, $\beta = 10^{-3}$, $\beta = 10^{-2}$, $\beta = 10^{-1}$, and $\beta = 1$. Figure 4.11(b) is the corresponding plot of inverse delay time as a function of the step current. This shows that $I_{step} \propto 1/t_d$ for current $I_{step} > I_{th}$. The dip in $1/t_d$ has a minimum value at the laser threshold current that separates a sustained lasing state with an injection current above I_{th} and the non-lasing state of the system that has an injection current below

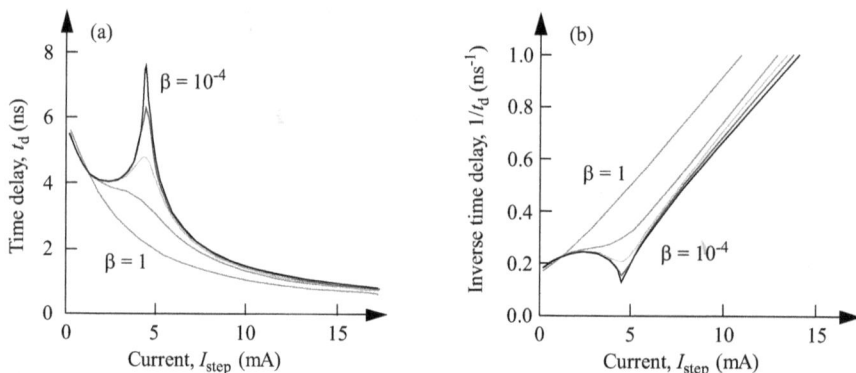

Figure 4.11. (a) Time delay t_d as a function of off–on step current, I_{step}, for values of spontaneous emission factor $\beta = 10^{-4}$, $\beta = 10^{-3}$, $\beta = 10^{-2}$, $\beta = 10^{-1}$, and $\beta = 1$. The macroscopic semiconductor laser diode has a threshold current, $I_{th} = 4.6$ mA, off current $I_{inj} = 0$ mA, and on current $I_{inj} = I_{step}$. (b) Inverse time delay, $1/t_d$, as a function of I_{step} for the values of spontaneous emission factor, β, in (a). The parameters in the calculations are those for an InGaAsP device given in table 4.1 with emission at the 1310 nm wavelength, $r_1 = 0.32$, and $r_2 = 0.98$.

I_{th}. Reduction of β acts to effectively decouple the lasing and non-lasing states of the system and so increases the value of t_d at the laser threshold. An increase in β introduces more spontaneous emission noise into the lasing mode and damps (reduces) the abruptness of the current–light lasing threshold characteristic. Increasing β reduces t_d and increases $1/t_d$. The value of $1/t_d$ at the laser threshold is proportional to an effective energy gap.

4.3.5 Cavity formation

The continuum mean-field single-mode rate equations (4.1) and (4.2) may be used to learn a great deal about the high-speed performance of a laser diode. The large-signal response to a step change in current predicts that turn-on delay and relaxation oscillations impact the switching speed of laser diodes. However, to some extent, these effects can be mitigated by engineering and design. With this in mind, it seems appropriate to ask if there are other, more fundamental, limitations to laser diode operation and switching speed. Of course, there are, and one such example is *cavity formation.*

To explain this, consider a photon inside a Fabry–Perot laser diode. The photon cannot know it is in a Fabry–Perot resonator or on resonance until it interacts with the mirrors. If there are no mirrors, the device is merely an LED with spectrally broadband spontaneous emission of light or, possibly, single-pass amplified spontaneous emission of light. Hence, lasing emission into a defined cavity resonance requires that the photon experiences at least one round-trip within the resonator. The device cannot behave as a laser until the photon cavity has formed. This takes *at least* one photon round-trip time, which, in a conventional Fabry–Perot laser diode, is about 10 ps.

The effect of multiple photon round-trips on the transient time-evolution of lasing light emission intensity and spectra involves fundamentals of device performance. It is not particularly easy to address because photon cavity formation is usually obscured by the nonlinear coupling of the optical field with the optical gain medium. Under normal conditions, it is difficult to measure the effect of multiple round-trips on the evolution of lasing light intensity and lasing spectra due to the short cavity round-trip time and charge carrier lifetime. However, by adiabatically decoupling the cavity formation from other processes, such as charge carrier dynamics, experiments can be performed to explore this issue [15]. Adiabatic decoupling may be realized experimentally by making a large external cavity. The results can also be predicted using time-delayed, single-mode or multi-mode rate equations. Importantly, the experiments show how a laser uses cavity formation to drive the device from an LED to a laser diode. Surprisingly, approximately $n_{RT} = 200$ photon round-trips are needed to approach steady-state $L_{out} - I_{inj}$ laser characteristics. The value of n_{RT} can be used to label discrete cavity formation states that asymptotically approach steady-state lasing behavior as n_{RT} increases (see figure 4.12). Obtaining pure steady-state *spectral* behavior requires even more photon round-trips.

Continuum mean-field single-mode time-delayed rate equations can be used to model transient optical intensity build-up in a laser with a passive external cavity. The inset in figure 4.13(a) shows the device configuration consisting of a semi-conductor diode gain medium coupled to an external fiber cavity with a photon

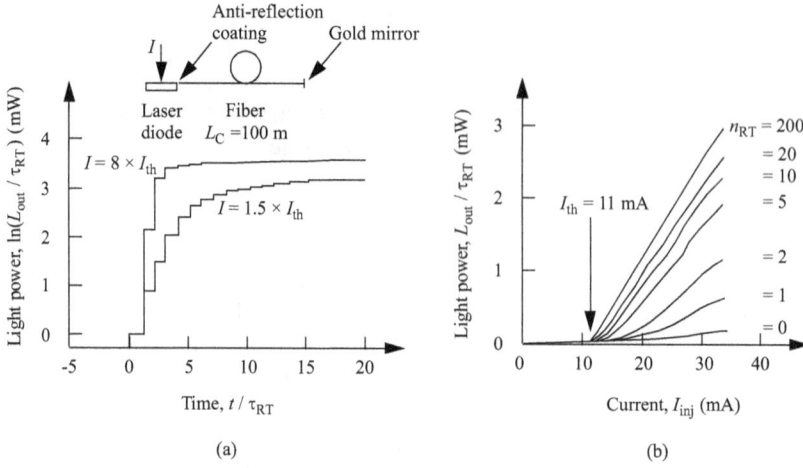

Figure 4.12. (a) Natural logarithm of laser emission intensity as a function of time, t/τ_{RT}. Time is measured in units of the cavity round-trip time $\tau_{RT} = 0.995\ \mu s$. The light level is normalized to the emission intensity when $0 < n_{RT} < 1$, where n_{RT} is the round-trip number in the cavity. The inset shows the experimental arrangement. The laser diode is the active region of a 100 m long Fabry–Perot resonator formed using a glass fiber wound on a spool. The diode is pumped by a step change in current, I_{inj}, and the light output power, L_{out}, is monitored as a function of the cavity round-trip time using a high-speed photodetector. (b) Normalized light output power as a function of injection current for the indicated number of photon round-trip trips, n_{RT}. When $n_{RT} = 0$, the device behaves as an LED. When $n_{RT} = 200$, the current–light output characteristics approach the steady-state laser diode behavior. (Adapted from [15], with the permission of AIP Publishing.)

round-trip time of $\tau_{RT} = 10$ ns. At time $t = 0$, there is a step change in injection current from $I_{inj} = 0$ mA to $I_{inj} = 30$ mA. For time $t > \tau_{RT}$, a fraction of the photons emitted into the external cavity will be reintroduced into the semiconductor gain medium, and some will be reflected from mirror 2 and remain in the external cavity. To take into account photons delayed by an integer number of round-trips, n_{RT}, before being reinjected into the active gain medium, the single-mode rate equations can be modified to

$$\frac{dS}{dt} = (G - \kappa)S - \beta r_{spon} + \kappa_2 \left(\sum_{j=1}^{n_{RT}} (1 - r_2) r_2^{(j-1)} r_3^j S(t - \tau_{RT}) \right) \Theta(t - \tau_{RT}), \quad (4.21)$$

where Θ is the step function, S is the photon density in the semiconductor gain medium, r_3 is the mirror reflectivity at the far right of the external cavity, and the rate of photon loss from mirror 2 is given by

$$\kappa_2 = \frac{1 - r_2}{2 - r_1 - r_2} \left(\frac{c}{2Ln_r} \right) \ln\left(\frac{1}{r_1 r_2} \right). \quad (4.22)$$

In this expression, the reflectivity of mirrors 1 and 2 are r_1 and r_2, respectively. If it is assumed that reflectivity at mirror 2 is very small, i.e. $r_2 \ll 1$, then only the first term in the sum in equation (4.21) needs to be retained, and the delayed rate equation for photon density becomes

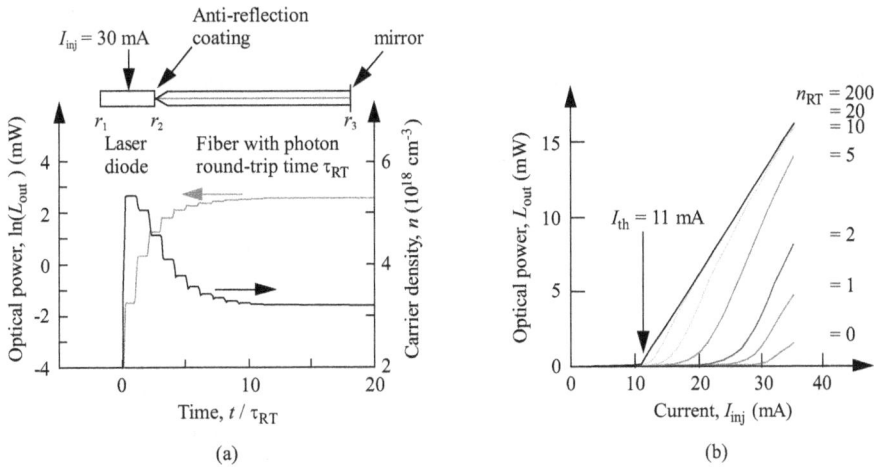

Figure 4.13. (a) Natural logarithm of calculated laser emission intensity, L_{out}, from the mirror with reflectivity r_1 as a function of time τ_{RT}. Time is measured in units of the external cavity round-trip time $\tau_{RT} = 10$ ns. Light output power, L_{out}, is monitored as a function of the cavity round-trip time using a high-speed photodetector. Calculated carrier density, n, as a function of time is also shown in the figure. Inset shows the device configuration and laser diode with a step change in injection current from $I_{inj} = 0$ mA from to $I_{inj} = 30$ mA. (b) Calculated light output power as a function of injection current for the indicated number of photon round-trip trips, n_{RT}. When $n_{RT} = 0$, the device behaves as an LED. When $n_{RT} = 200$, the current–light output characteristics approach the steady-state behavior of a laser diode with threshold current $I_{th} = 11$ mA. The parameters are those of the laser diode with emission wavelength $\lambda_0 = 1310$ nm given in table 4.1 with mirror reflectivity $r_1 = 0.32$, $r_2 = 2 \times 10^{-6}$, and $r_3 = 1$.

$$\frac{dS}{dt} = (G - \kappa)S - \beta r_{spon} - \kappa_2(1 - r_2)r_3 S(t - \tau_{RT}). \qquad (4.23)$$

Figure 4.13(a) shows calculated laser emission intensity as a function of time, τ_{RT}, using this approximation. Also shown is the calculated carrier density as a function of time. Both optical emission and carrier density have a time dependence that is determined stepwise by the external cavity round-trip time $\tau_{RT} = 10$ ns.

Figure 4.13 (b) shows results of calculating optical output power as a function of injection current, I_{inj}, for the indicated number of photon round-trip trips, n_{RT}. When $n_{RT} = 0$, the device behaves as an LED. When $n_{RT} = 200$, the current–light output characteristics approach the steady-state behavior of a laser diode with threshold current $I_{th} = 11$ mA. The qualitative agreement between the calculations shown in figure 4.13 (b) and the experimental results shown in figure 4.12 serves to validate the approach taken to model the device.

4.4 Small-signal intensity response

The small-signal intensity response is studied by considering the effect of a small sinusoidal modulation in current of amplitude ΔI_{inj} about some average value $\langle I_{inj} \rangle$ has on laser light emission intensity. Current is

$$I_{\text{inj}}(t) = \langle I_{\text{inj}} \rangle + \text{Re}(\Delta I_{\text{inj}} e^{i\omega_m t}), \qquad (4.24)$$

where ω_m is the small-signal modulation frequency. Because it is assumed $\Delta I \ll \langle I_{\text{inj}} \rangle$ the rate equations can be linearized and the variation in carrier density, Δn, and photon density, ΔS, is also sinusoidal at the same frequency, ω_m. Hence,

$$n(t) = \langle n \rangle + \text{Re}(\Delta n e^{i\omega_m t}) \qquad (4.25)$$

and

$$S(t) = \langle S \rangle + \text{Re}(\Delta S e^{i\omega_m t}). \qquad (4.26)$$

Substituting into the laser rate equations (equations (4.1) and (4.2)) and, for a conventional laser diode with $\beta \ll 1$, ignoring the small contribution of spontaneous emission into the lasing mode gives the small-signal transfer function that relates change in photon density to change in current for $I_{\text{inj}} > I_{\text{th}}$. The small-signal response has a characteristic resonant relaxation frequency [16]

$$\omega_r = \sqrt{\frac{1}{\tau_{\text{ph}} \tau_{\text{n}}} \left(\frac{I_{\text{inj}}}{I_{\text{th}}} - 1 \right)} \qquad (4.27)$$

that increases as the square root of $I_{\text{inj}} - I_{\text{th}}$. The characteristic temporal decay of the resonance is of the form $e^{-t/\tau_{\text{damp}}}$ in which the damping rate $\gamma_r = 1/\tau_{\text{damp}}$, where

$$\gamma_r = \frac{1}{2} \left(\frac{1}{\tau_{\text{n}}} + \tau_{\text{ph}} \omega_r^2 \right). \qquad (4.28)$$

A portion of this chapter was reproduced with permission from [17].

Bibliography

[1] Runge C 1895 *Math. Ann.* **46** 167
[2] Kutta W 1901 *Z. Math. Phys.* **46** 435
[3] Press W H, Teukolsky S A, Vetterling W T and Flannery B P 2007 *Numerical Recipes: The Art of Scientific Computing* 3rd edn (Cambridge: Cambridge University Press)
[4] For a review, see Vasil'ev P P, White I H and Gowar J 2000 *Rep. Prog. Phys.* **63** 1997
[5] O'Gorman J, Levi A F J, Tanbun-Ek T and Logan R A 1991 *Appl. Phys. Lett.* **59** 16
[6] Andrekson P A, O'Gorman J, Levi A F J, Haner M, Olsson N A, Tanbun-Ek T, Coblentz D L and Logan R A 1991 *IEEE Trans. Photon. Technol. Lett.* **3** 1150
[7] Levi A F J, Nottenburg R N, Nordin R A, Tanbun-Ek T and Logan R A 1990 *Appl. Phys. Lett.* **56** 1095
[8] Hall J L and Ye J 2003 *IEEE Trans. Instr. Measure.* **52** 227
[9] For a review, see Chang L, Liu S and Bowers J E 2022 *Nat. Photon.* **16** 95
[10] Kudelin I *et al* 2024 *Nature* **627** 534
[11] Sun S *et al* 2024 *Nature* **627** 540
[12] He Y, Cheng L, Wang H, Zhang Y, Meade R, Vahala K, Zhang M and Li J 2024 *Sci. Adv.* **10** eado9570
[13] DeGiorgio V and Scully M O 1970 *Phys. Rev.* A2 *1170*

[14] Haken H 1975 *Rev. Mod. Phys.* **47** 67
[15] O'Gorman J, Levi A F J, Coblentz D, Tanbun-Ek T and Logan R A 1992 *Appl. Phys. Lett.* **61** 889
[16] Adams. M J 1973 *Opto-Electron.* **5** 201
[17] Levi A F J 2023 *Applied Quantum Mechanics* 3rd edn (Cambridge: Cambridge University Press)

IOP Publishing

Essential Semiconductor Laser Device Physics (Second Edition)

A F J Levi

Chapter 5

Noise and fluctuations

This chapter describes the relative intensity noise and bit error ratio of an optical non-return to zero digital signal as a function of signal-to-noise ratio. The shot-noise limit to relative intensity noise using a perfect photodetector is introduced as is the existence of squeezed states. The basics of numerical simulation of relative intensity noise using the Langevin rate equations are provided, as well as the phase noise Langevin rate equations and the linewidth enhancement factor in semiconductor diode lasers. The modified Schawlow–Townes semiconductor laser diode emission spectral linewidth is discussed. The temperature dependence of semiconductor laser diode threshold current and the key role played by Landau–Ginzburg second-order phase-transition fluctuations are described.

A laser works by amplifying spontaneous emission, which is fundamentally quantum-mechanical and random. These random photon emission events contribute to noise in emitted laser light, especially at low light levels. Because fiber-optic communication systems often use intensity modulation of laser light to transmit information, there is interest in understanding intensity noise and other noise sources. A typical fiber-optic link converts an electrical source of binary data to a time-modulated light output that is transmitted through a low-loss fiber medium, received, and converted back to an electrical data stream.

Figure 5.1(a) shows how a non-return to zero (NRZ) intensity-modulated laser light output, $L_{\text{out}}(t)$, can be used to transmit classical bits of information in a digital data link. In this example, a high level of light is binary 1 and a low level of light is binary 0. In a balanced system, the average light level is midway between the high and low levels and so the average bit value is 1/2. Because the receiver circuitry needs time to decide between high and low light levels, in conventional systems, data are transmitted at a well-defined rate called the *bit rate*, $1/\tau_{\text{bit}}$. A typical bit rate in fiber-optic communication systems is 25 Gb s^{-1}. In this case, $\tau_{\text{bit}} = 40$ ps and 25×10^9 bits (25 Gb) of information can be transmitted through the system in one second. As with any information channel, it is important to minimize bit errors due to the presence of

doi:10.1088/978-0-7503-6417-1ch5

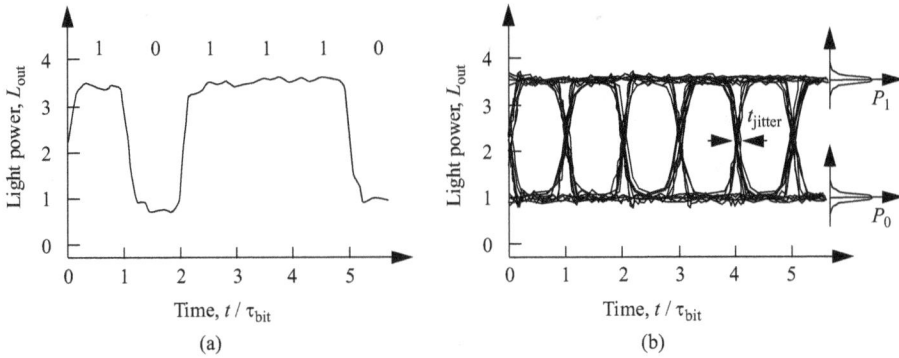

Figure 5.1. (a) Intensity-modulated laser light can transmit bits of information in which a high level of light is binary 1, and a low level of light is binary 0. Signal intensity noise and timing jitter are sources of bit errors. (b) A continuously sampled data stream creates an eye diagram in which the eye width and height define the area where the receiver circuitry can minimize the probability of bit errors. Noise creates a statistical distribution for the low and high light levels, shown schematically as P_0 and P_1. There is also a distribution for timing jitter, t_{jitter}.

noise. The bit error *ratio* (BER) is the number of errors divided by the number of transmitted bits so that a BER of 4×10^{-11} in a system transmitting at 25 Gb s^{-1} will, on average, produce one error per second. Notice that the bit error *ratio* is dimensionless (in particular, it is *not* a bit error rate which, of course, would have units of inverse seconds and not be a useful measure in system design).

Continuous sampling of a time-domain NRZ random binary data stream referenced to a bit rate clock generates the *eye diagram* shown in figure 5.1(b). The opening near the center of the eye diagram is where a receiver comparator circuit decides between high and low light levels. The comparator threshold value is optimally set to minimize errors. The greater the amount of intensity noise and timing jitter in the optical signal the greater the probability of errors.

If, as is usually the case, the receiver uses a reverse-biased p–i–n photodiode to detect light then thermal noise, which is independent of the signal current, can be the dominant source of intensity noise. In this situation, noise in the high and low levels of the NRZ signal is the same and, as shown schematically in figure 5.1(b), is normally distributed as P_1 and P_0, respectively. In this case, the optimum intensity threshold value for the receiver comparator is at the midpoint between the average binary high value and average binary low value. Ignoring the contribution of timing jitter, the ratio of signal power to noise (SNR) is related to BER by

$$\text{BER} = \frac{1}{2}\left(1 - \text{erf}\left(\frac{\sqrt{\text{SNR}}}{2\sqrt{2}}\right)\right), \tag{5.1}$$

where erf is the error function. As indicated in figure 5.2, BER is a rapidly varying function when SNR $\geqslant 20$. In this example, an SNR of 169 (22.28 dB) is required to achieve a BER of 4×10^{-11}.

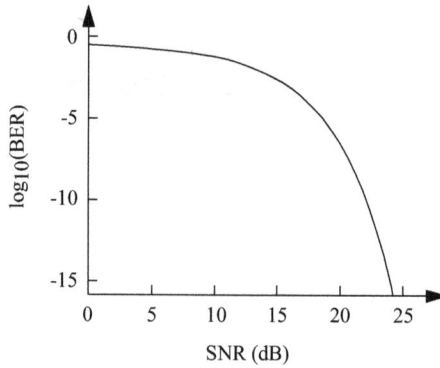

Figure 5.2. BER as function of SNR for an NRZ signal dominated by p–i–n thermal noise in the optical receiver.

Thermal noise is not the only source of fluctuations contributing to noise in the received signal. Additional sources include relative intensity noise, shot noise, phase noise, and jitter. Data can be transmitted in modulation formats other than NRZ and so have different sensitivity to error creation in the presence of noise. In addition, forward error correction (FEC) coding can be used to recover from errors generated in a fiber-optic link.

5.1 Relative intensity noise

Fluctuation in the intensity of light output from a laser diode is characterized by relative intensity noise (RIN). The optical intensity S contains noise, so that

$$S(t) = \langle S \rangle + \delta S(t), \tag{5.2}$$

where $\langle S \rangle$ is the time-averaged optical intensity and $\delta S(t)$ is the deviation from the average value at any given instant in time, t. The time average of the fluctuation is $\langle \delta S(t) \rangle = 0$. The noise $\delta S(t)$ may be characterized in the time domain by the autocorrelation function

$$g_S(\tau) = \langle \delta S(t)\delta S(t - \tau) \rangle \tag{5.3}$$

or in the frequency domain by the noise spectral density

$$\langle |\delta S(\omega)|^2 \rangle = \int_{\tau=-\infty}^{\tau=\infty} g_S(\tau)\mathrm{e}^{-\mathrm{i}\omega\tau}\mathrm{d}\tau = \frac{1}{t'}\bigg|_{t'\to\infty} \left| \int_0^{t'} \delta S(t)\mathrm{e}^{-\mathrm{i}\omega t}\mathrm{d}t \right|^2. \tag{5.4}$$

RIN is defined [1] as the square of fluctuations in optical intensity noise at frequency ω divided by the average value of optical intensity $\langle S \rangle = S_0$ squared, so that

$$\mathrm{RIN}(\omega) = \frac{\langle |\delta S(\omega)|^2 \rangle}{\langle S \rangle^2} = \frac{\langle |\delta S(\omega)|^2 \rangle}{S_0^2}. \tag{5.5}$$

RIN is measured either in units of dBc Hz^{-1} or Hz^{-1}. RIN can be as low as -165 dBc Hz^{-1} in a good laser diode and an RIN of -110 dBc Hz^{-1} makes an insignificant contribution to a BER of 10^{-9}.

A high-speed photodetector converts photons to electrical current, $i(t)$. Photon flux is proportional to electric current and so the fluctuation $\delta S(t) \propto i(t)$. Because electrical RF *power* across load resistor R_L is $|i(\omega)|^2 R_L$ it follows that the average noise power spectral density per unit frequency in the photodetector electrical current is

$$\delta P(\omega) = \langle |i(\omega)|^2 \rangle R_L \propto \langle |\delta S(\omega)|^2 \rangle \tag{5.6}$$

and the average power in the electrical current is

$$\langle P \rangle = P_0 \propto \langle S \rangle^2. \tag{5.7}$$

Hence, RIN measured using a perfect photodetector and a perfect electrical RF spectrum analyzer is simply

$$\text{RIN}(\omega) = \frac{\delta P(\omega)}{P_0}. \tag{5.8}$$

5.1.1 Shot-noise limit to RIN

The relative intensity noise due to the discrete quantum nature of the photon gives rise to a shot-noise contribution to the RIN. A time-average optical power P_0 detected by the photodetector consists of a flux of photons of energy $\hbar\omega_{ph}$ with a shot-noise RIN of

$$\text{RIN}_{\text{shot-noise}} = \frac{2\hbar\omega_{ph}}{P_0}. \tag{5.9}$$

The factor of two takes into account both positive and negative frequency contributions because it is assumed the photodetector is attached to an RF spectrum analyzer that measures a single-sided frequency spectrum.

For example, $P_0 = 1$ mW laser light of wavelength $\lambda = 1310$ nm has

$$\text{RIN}_{\text{shot-noise}} = \frac{2 \times 1.05 \times 10^{-34} \times 3 \times 10^8 \times 2\pi}{10^{-3} \times 1310 \times 10^{-9}} \tag{5.10}$$
$$= 3 \times 10^{-16} \text{ Hz}^{-1} = -155 \text{ dBc Hz}^{-1}.$$

Increasing P_0 reduces $\text{RIN}_{\text{shot-noise}}$. Assuming a broad (white) spectrum for shot noise, $\text{RIN}_{\text{shot-noise}}$ is, absent quantum-mechanical squeezing, a noise floor.

5.1.2 Squeezed states

The uncertainty relation in quantum mechanics requires that the standard deviation in photon number Δs and phase $\Delta \theta$ of a state satisfies the inequality [2]

$$\Delta s \Delta \theta \geqslant \frac{1}{2}. \tag{5.11}$$

In a standard *coherent state*, the uncertainty is minimized with equal contributions from the two components Δs and $\Delta \theta$. The more general *squeezed* coherent state minimizes uncertainty with unequal contributions from the two components [3]. Hence, of practical interest is the prospect of creating systems that exploit the fact that photon number fluctuations Δs can be reduced to a value less than that in a standard coherent state.

In principle, it is possible to suppress photon number fluctuations inside the optical cavity of a laser diode at high injection currents via nonlinear optical processes. For example, if the gain saturation coefficient in equations (4.5) or (4.6) is modified such that optical gain decreases rapidly with increasing photon number, s, then the resulting negative feedback in the system can efficiently reduce photon number fluctuations. Any value of Δs smaller than that predicted for s photons in a standard coherent state is at the expense of increasing photon phase uncertainty, $\Delta \theta$. Such *squeezing* of a coherent state is a quantum-mechanical effect. It may be possible to engineer a laser diode to enhance nonlinear optical gain behavior and create a super gain saturation mechanism that efficiently creates squeezed states and damps relaxation oscillations [4]. Separately, quantum dot laser diodes driven by a low noise current source can also be configured to exhibit photon output intensity noise below the classical shot-noise limit [5]. However, any device design should take into consideration photon loss and noise mechanisms that can reduce squeezing in the emitted photon flux.

Squeezed coherent photon sources have applications in sensing, communication, information processing, and metrology [6].

5.2 Langevin intensity rate equations

The RIN of photon emission from a laser diode may be investigated numerically by changing the single-mode rate equations to *Langevin* rate equations,

$$\frac{dn}{dt} = \frac{I_{inj}}{e V_{vol}} - \frac{n}{\tau_n} - GS + F_n(t) \tag{5.12}$$

and

$$\frac{dS}{dt} = (G - \kappa)S + \beta r_{spon} + F_s(t), \tag{5.13}$$

where n and S are the carrier and photon density in the cavity, respectively, G is optical gain, κ is optical loss, β is the spontaneous emission factor, n/τ_n is the carrier recombination rate, e is the charge of an electron, r_{spon} accounts for spontaneous emission into all optical modes, and V_{vol} is the volume of the semiconductor active region. A source of random noise in the rate equations is included through the terms $F_n(t)$ and $F_s(t)$. These are the Langevin noise terms. Equations (5.12) and (5.13) are the simplest way to include noise in a laser diode model. A slightly more detailed approach also takes into account optical phase noise [7]. In the Markovian approximation, corresponding to instantaneous changes in $F_n(t)$ and $F_s(t)$, the autocorrelation and cross-correlation functions are given by

$$\langle F_n(t)F_n(t')\rangle = \left(\frac{I_{inj}}{eV_{vol}} + GS + \frac{n}{\tau_n}\right)\delta(t - t') \tag{5.14}$$

$$\langle F(t)_s F(t')_s\rangle = ((G + \kappa)S + \beta r_{spon})\delta(t - t') \tag{5.15}$$

$$\langle F_s(t)F_n(t')\rangle = -(GS + \beta r_{spon})\delta(t - t'). \tag{5.16}$$

The Markovian approximation is guaranteed by the use of $\delta(t - t')$. $\langle F_n(t)F_n(t')\rangle$ is the square of Gaussian fluctuations around the mean value of n given by continuum mean-field rate equations and $\langle F(t)_s F(t')_s\rangle$ is the square of Gaussian fluctuations around the mean value of S. The cross-correlation term $\langle F_s(t)F_n(t')\rangle$ shows that the rate equations for S and n are coupled and hence correlated. The negative sign in equation (5.16) indicates that $F_n(t)$ and $F_s(t)$ are anti-correlated.

Numerical integration of Langevin rate equations (5.12) and (5.13) using the fourth-order Runge–Kutta method with a uniform time step t_{step} introduces noise that is simulated using a normalized Gaussian random number generator (randn in MATLAB). The noise term for electron density is

$$F_n(t) = \text{randn} \times \sqrt{\frac{\frac{I_{inj}}{eV_{vol}} + GS + \frac{n}{\tau_n}}{t_{step}}} \tag{5.17}$$

and the noise term for photon density is

$$F_s(t) = \text{randn} \times \sqrt{\frac{(G + \kappa)S + \beta r_{spon}}{t_{step}}}. \tag{5.18}$$

The mean steady-state value for n is $\langle n\rangle$ and that for S is $\langle S\rangle$. The magnitude of the Fourier component of n at RF angular frequency ω is $\delta n(\omega)$ and that for S is $\delta S(\omega)$. Linearizing the dynamical equations (5.12) and (5.13) for a constant average diode current, and neglecting gain saturation, gives

$$\delta S(\omega) = \frac{F_n(\omega)\left(\frac{dG}{dn}\langle S\rangle + 2\beta B\langle n\rangle\right) + F_s(\omega)\left(i\omega + \frac{dG}{dn}\langle S\rangle + \frac{n}{\tau_n} + \frac{1}{\tau_n'}\langle n\rangle\right)}{\left(i\omega\left(\frac{dG}{dn}\langle S\rangle + \frac{n}{\tau_n} + \frac{1}{\tau_n'}\langle n\rangle\right) - \omega^2 + G\frac{dG}{dn}\langle S\rangle + G2\beta B\langle n\rangle\right)} \tag{5.19}$$

and

$$\text{RIN} \equiv \lim(\tau \to \infty)\frac{1}{\tau}\left|\frac{\delta S(\omega)}{\langle S(\omega)\rangle}\right|^2. \tag{5.20}$$

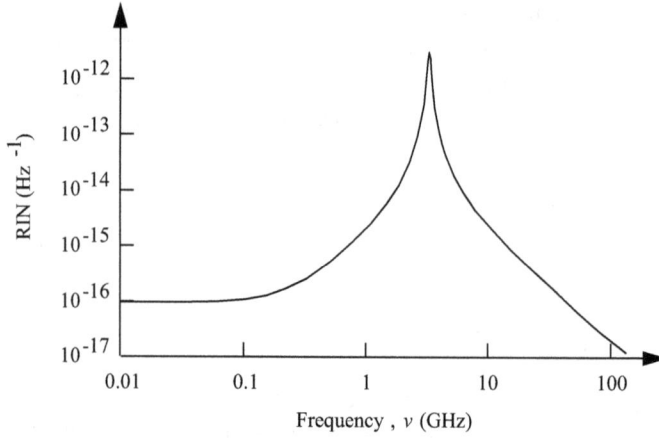

Figure 5.3. Calculated RIN as a function of frequency, ν, for a Fabry–Perot laser diode with active volume $V_{\text{vol}} = 300 \times 2 \times 0.05\ \mu\text{m}^3$. The mirror reflectivity is 0.3 per facet and the steady-state current bias is $I_0 = 4 \times I_{\text{th}} = 7.36\ \text{mA}$. The average photon number in the device is 9.5×10^4 and the average carrier number is 5.9×10^7.

In these expressions, $F_n(\omega)$ and $F_s(\omega)$ are the Fourier components at ω of $F_n(t)$ and $F_s(t)$, respectively, and $1/\tau'_n \equiv (\text{d}/\text{d}n)(1/\tau_n)$.

The cause of the peak in RIN shown in figure 5.3 can be traced to the pole in equation (5.19). Physically, it arises from fluctuations in carrier density and photon density, which work in phase to amplify the response to noise fluctuation. The origin is similar to the cause of relaxation oscillation observed in the average large-signal response of carriers and photons.

5.2.1 Phase noise Langevin rate equations and linewidth enhancement factor

A plane-wave electromagnetic field with photon wave number k_{ph} propagating in the z-direction in a homogeneous isotropic linear gain medium characterized by complex refractive index

$$n_{\text{index}} = n_r + i\kappa_r \tag{5.21}$$

can be described as

$$E_{\text{opt}} = E_0 e^{ik_{\text{ph}} n_{\text{index}} z} = E_0 e^{ik_{\text{ph}} n_r z} e^{-k_{\text{ph}} \kappa_r z}, \tag{5.22}$$

where $k_{\text{ph}} = 2\pi/\lambda$. Power flux is proportional to $|E_{\text{opt}}|^2$ and increases with distance z if the imaginary part of the refractive index κ_r is *negative* since

$$\frac{\text{d}P_{\text{flux}}}{\text{d}z} = -2k_{\text{ph}} \kappa_r P_{\text{flux}}(z). \tag{5.23}$$

It follows that optical gain in the lasing mode is related to the imaginary part of the refractive index κ_r via

$$G = -2k_{\text{ph}} \kappa_r. \tag{5.24}$$

Changes in carrier density change the optical gain and κ_r. Because the imaginary and real parts of the refractive index are related by the Kramers–Kronig relations, any change in the imaginary part of the refractive index $\delta\kappa_r$ also changes the real part of the refractive index by an amount δn_r. The ratio

$$\alpha = \frac{\delta n_r}{\delta \kappa_r} \tag{5.25}$$

is called the *linewidth enhancement factor* and typically has a value in the range from 2 to 8 in a conventional laser diode with a bulk active region, indicating that in a bulk semiconductor n_r is more strongly influenced by stimulated emission than κ_r.

Making use of this fact, the gain can be related to n_r via α giving the single-mode rate equation for phase

$$\frac{d\phi}{dt} = \frac{\alpha \Gamma v_g G_{\text{slope}}}{2}(n - \langle n \rangle) + F_\phi(t), \tag{5.26}$$

where v_g is group velocity and

$$F_\phi(t) = \text{randn} \times \sqrt{\frac{(G + \kappa)S + \beta r_{\text{spon}}}{4S^2 t_{\text{step}}}}. \tag{5.27}$$

The value of the linewidth enhancement factor depends on the semiconductor active region used. A quantum well structure can operate with a value of α that is less than that of the bulk active material [8] and a semiconductor quantum dot active region can have a value of α that is much less than unity [9].

5.2.2 Spectral linewidth

The full-width-at-half-maximum (FWHM) of laser light emission spectral linewidth from a conventional laser diode can be shown to be

$$\Delta \nu_{\text{FWHM}} = \frac{(\alpha^2 + 1)\beta r_{\text{spon}}}{4\pi \tau_{\text{ph}} \langle S \rangle}, \tag{5.28}$$

which is the modified Schawlow–Townes formula [10–14]. This illustrates the importance of the linewidth enhancement factor, spontaneous emission, photon lifetime, and photon intensity in determining spectral linewidth. The equation predicts that spectral linewidth can be reduced by operating the device at high photon intensity in a cavity with long photon lifetime. However, at high photon intensity in the laser cavity, gain saturation can reduce carrier pinning and this can lead to an increase in $\Delta\nu_{\text{FWHM}}$ with increasing $\langle S \rangle$. Gain saturation can have contributions from spectral and spatial hole burning. One consequence of carrier de-pinning is an increase in the spontaneous emission term, r_{spon}, with increasing injection current. Also, at sufficiently high photon intensity, not only does gain compression reduce efficiency in the conversion of current into photons, it also increases the value of β because the spectral shape of κ can contact a finite portion of the gain spectrum.

The fact that $\Delta\nu_{\text{FWHM}}$ scales as $1/\langle S \rangle$ in equation (5.27) can be explained by considering the limiting case in which many photons are present in the lasing mode. In this situation, the electromagnetic field is classical with an instantaneous value $\mathbf{E}(t)$ that is proportional to $\sqrt{S(t)}\,e^{-i\phi(t)}$, where $S(t)$ has a Gaussian distribution about the mean value $\langle S \rangle$ and the phase ϕ changes randomly in the range $0 \leqslant \phi < 2\pi$ due to spontaneous emission. The contribution of fluctuations in phase $\delta\phi$ to laser spectral linewidth is proportional to the diffusion constant $D_\phi = \beta(\delta\phi)^2/\tau_{\text{spon}}$. If the random fluctuation in the electromagnetic field is due to spontaneous emission, then the corresponding fluctuation in phase $\delta\phi$ is proportional to $1/\sqrt{S}$ and time-averaged laser linewidth $\Delta\nu_{\text{FWHM}} \propto 1/\langle S \rangle$.

5.3 Temperature dependence of the semiconductor laser diode threshold current

Injection of current into a forward-biased light-emitting diode (LED) results in spontaneous emission of photons contributing to a spectrally broad range of optical emission frequencies. In contrast, photons from a properly designed single-mode laser diode operating with an injection current above the laser threshold value, I_{th}, are emitted into a spectrally narrow range of optical emission frequencies. However, a laser diode with a forward-biased injection current $I_{\text{inj}} \ll I_{\text{th}}$ has an optical emission spectrum similar to that of an LED. As a starting point to understand the temperature dependence of the transition between these operating states in a conventional macroscopic Fabry–Perot laser diode, it is convenient to assume that optical emission consists of separate independent lasing and nonlasing components. When this is done at a fixed injection current, it is found that the integrated broadband emission of the nonlasing component depends approximately exponentially on device temperature over an appropriate temperature range, while the integrated narrow-band lasing emission component has a power law dependence below laser threshold. Experiments also show that net optical gain is more important than nonradiative recombination in determining the temperature-dependent characteristics of long-wavelength semiconductor laser diodes.

Figure 5.4 is a plot of measured light output power from a semiconductor laser diode as a function of temperature for fixed injection current, $I_{\text{inj}} = 10.5$ mA. The total optical output does not depend exponentially on temperature near $T = 29\,°\text{C}$ (302 K), below-which lasing occurs.

The device used for the measured results shown in figure 5.4 has a laser threshold current of $I_{\text{th}} = 8.5$ mA at a temperature of $T = 20\,°\text{C}$ (293 K). While figure 5.4 shows that the total optical output does not depend exponentially on temperature, the laser diode *threshold current*, I_{th}, as a function of temperature between 10 °C and 70 °C (283 K and 343 K), has a best-fit characteristic temperature $T_0^{\text{LD}} = 42$ K in the expression $I_{\text{th}} = I_0 e^{T/T_0^{\text{LD}}}$.

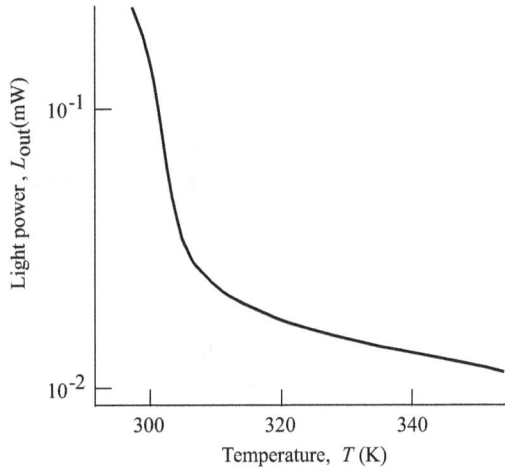

Figure 5.4. Semilogarithmic plot of measured total light output power of a Fabry–Perot semiconductor laser diode with an emission wavelength near $\lambda = 1300$ nm as a function of substrate temperature for a constant injection current, $I_{inj} = 10.5$ mA. At this fixed injection current, lasing occurs at temperatures below $T = 29\,^{\circ}\text{C}$ (302 K). The device has a laser threshold current of $I_{th} = 8.5$ mA at a temperature of $T = 20\,^{\circ}\text{C}$ (293 K). (Adapted with permission from [21].)

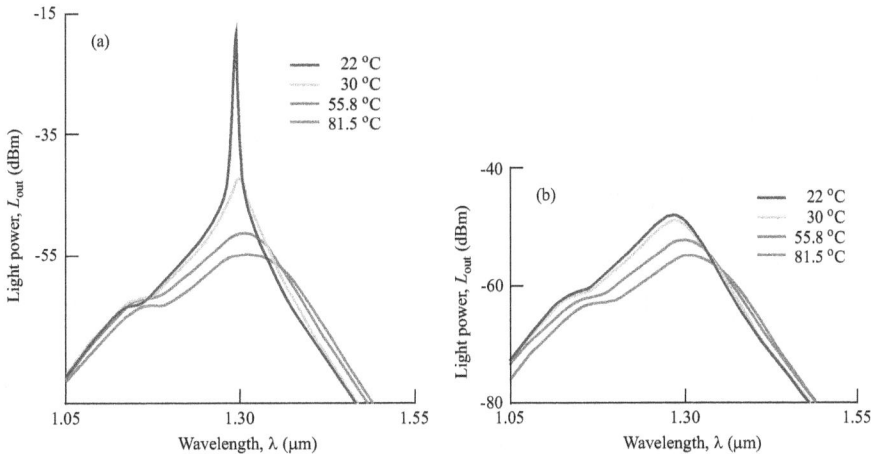

Figure 5.5. (a) Measured spectra of a Fabry–Perot semiconductor laser diode with emission wavelength near $\lambda = 1300$ nm for the indicated substrate temperatures and at constant injection current, $I_{inj} = 10.5$ mA. (b) Measured emission spectra of the same device as in (a) but with mirror reflectivity reduced to $r < 0.001$. The device behavior is similar to that of an LED. (Adapted with permission from [21].)

Figure 5.5(a) shows laser diode emission spectra for the indicated substrate temperatures, while figure 5.5(b) similarly shows the emission spectra of an LED at the same temperatures. The LED is the same device as the laser diode other than the addition of antireflection-coated mirrors ($r < 0.001$). A best-fit of the integrated LED output power $P_{LED}(T)$ to an exponential temperature dependence results in a characteristic temperature $T_0^{LED} = 115$ K. In this case, the ratio of T_0^{LED} and T_0^{LD} is 2.7. This is a significant difference because the characteristic temperature T_0 appears in the exponent.

At the relatively high temperatures of 55.8 °C and 81.5 °C, the result shown in figure 5.5 indicates no lasing and that the spectrally broad emission features of the laser diode and the LED are essentially identical. With decreasing device temperature, the semiconductor band-gap energy increases, causing broadband emission for both devices to peak at shorter wavelengths. As temperature is decreased further, becoming less than $T = 29$ °C (302 K), lasing occurs in the laser diode, and emitted optical power from the laser diode increases into the lasing wavelength. At the same time, the broadband emission portion of the laser diode spectrum does not change because of the carrier pinning associated with lasing.

The experimental results shown in figure 5.5 support the assumption that optical emission from a laser diode can be approximated in a model consisting of contributions from separate independent lasing and nonlasing components. This and its limits can be justified by considering the separation of temporal and spatial scales that can occur in more detailed physical models.

The existence of lasing and nonlasing components of a macroscopic laser diode with a threshold about which one component tends to dominate is suggestive of a

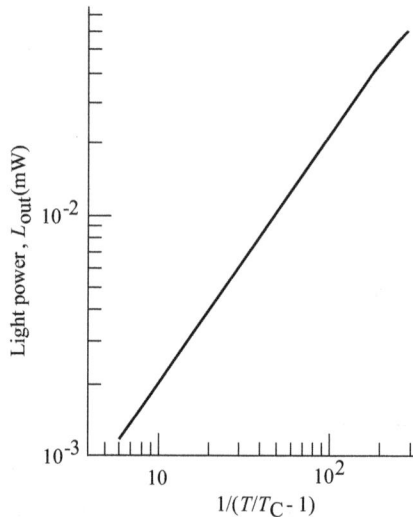

Figure 5.6. Logarithmic plot of measured lasing light output power, L_{out}, as a function of reduced temperature $T_C/(T - T_C)$ of a macroscopic Fabry–Perot semiconductor laser diode with emission wavelength near $\lambda = 1300$ nm, with $I_{inj} = 10.5$ mA, and $T_C = 301.4$ K. The lasing light component is obtained by subtraction of the measured temperature-dependent integrated LED output $P_{LED}(T)$ from the total measured light output of the laser diode. Laser light power near the laser threshold behaves according to the Landau–Ginzburg theory of second-order phase transitions. Deviation from linear behavior in the plot can occur as $T \to T_C$ (so that $(T_C/(T - T_C) \to \infty)$ and this is due to the increasing importance of nonlinear fluctuations, which act to remove the singularity at $T = T_C$. (Adapted with permission from [21].)

driven nonequilibrium phase transition. In fact, adopting the magnitude of laser electric field as the order parameter, optical emission around the laser threshold follows the Landau–Ginzburg theory of second-order phase transitions in which T_C is a critical temperature associated with the phase transition. In this description, below-threshold emitted light is in an incoherent, disordered state with unsustainable fluctuations into the ordered state, and above threshold, the emitted light is in a coherent, ordered, state. Below threshold $\langle S \rangle \sim |T/T_C - 1|^{-\gamma_C}$, where the critical exponent $\gamma_C = 1$ and the mean square fluctuation of the lasing field amplitude is the average lasing photon output, $\langle S \rangle$. Nonlinear fluctuations remove the singularity at $T = T_C$.

Figure 5.6 shows the measured lasing light component of a macroscopic semiconductor laser diode versus reduced temperature, $(T/T_C - 1)^{-1}$. The straight line in figure 5.6 over nearly two orders of magnitude is evidence confirming the predictions of the Landau–Ginzburg theory. The integrated narrow-band lasing emission has a power law dependence on temperature below the laser threshold, corresponding to unsustainable fluctuations into the lasing mode.

5.3.1 The role of Landau–Ginzburg phase-transition fluctuations in determining the laser diode threshold current

Spontaneous emission of photons is a random incoherent process whereas stimulated processes in a laser can result in emission of coherent light. Driving a macroscopic laser towards laser threshold results in optical emission whose statistical properties are analogous to those of second-order thermodynamic phase transitions [15–20]. Increasing the injection current of a semiconductor laser diode towards laser threshold results in fluctuations into a lasing state. While these fluctuations into the lasing state are unsustainable when the injection current is below the threshold current, they can efficiently reduce below-threshold carrier density. To understand this operating regime, it is useful to view optical emission from a laser diode as consisting of lasing and nonlasing components whose relative contribution varies. In a macroscopic Fabry–Perot laser diode the *nonlasing* component has spectrally broadband emission into nonlasing optical modes. This is similar to the emission spectrum of a conventional LED with the exception that single-pass amplified spontaneous emission is possible along the relatively long active region of the device. The *lasing* component of the system has spectrally narrow emission into the lasing mode. In direct analogy with the Landau–Ginzburg theory of second-order phase transitions [21], the lasing component has a temperature dependence below threshold that may be characterized by a power law.

A conventional macroscopic Fabry–Perot laser diode can be converted into an LED by antireflection coating the mirrors of the device. Using this approach, it is possible to compare the effect of subthreshold fluctuations in a laser diode with that of an LED that has the *same* geometry and active region [22, 23]. Experiments using this technique show that subthreshold fluctuations of photons into the optical modes of a laser diode remove a current, I_{fl}, whose relative importance in determining threshold current increases with temperature.

Carrier density, n, as a function of injection current, I_{inj}, is shown in figure 5.7 for an InP buried heterostructure Fabry–Perot laser diode and an LED that have the same geometry and bulk InGaAs active region. As may be seen in figure 5.7(a), when device temperature is $T = 298$ K, for a given injection current, I_{inj}, fluctuations into the lasing state of the laser diode reduce carrier density when compared to the LED. It is for this reason that an additional current, $I_{fl} = 4$ mA, is needed to reach laser threshold current, $I_{th} = 9.7$ mA, and threshold carrier density, n_{th}. The current I_{fl} associated with fluctuations into the lasing state is the difference in injection current between LED and laser diode to reach the laser diode threshold carrier density, n_{th}. Figure 5.7(b) shows that increasing device temperature to $T = 328$ K has the effect of increasing I_{th}, n_{th}, and the relative contribution of the current I_{fl}. As temperature is increased the value of n_{th} increases and the contribution of I_{fl} also increases. Subthreshold fluctuations into the lasing state are a critical part of a feedback mechanism that causes laser threshold current I_{th} to increase with increasing device temperature.

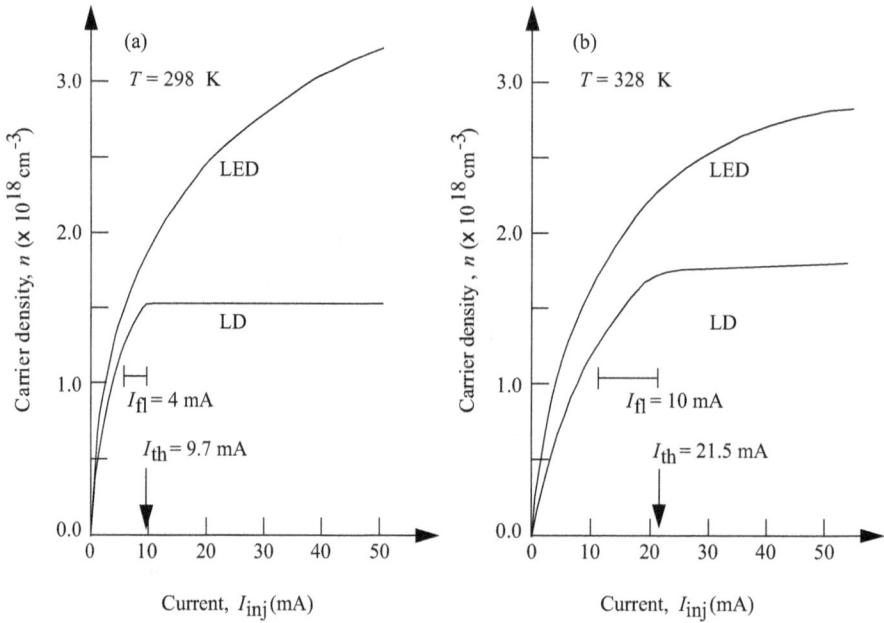

Figure 5.7. Plot of carrier density, n, as a function of injection current, I_{inj}, in an LED and Fabry–Perot laser diode that have the *same* geometry and active region. The InP buried heterostructure device has a bulk active InGaAs region with band-gap energy $E_g = 0.9686$ eV, thickness $t_a = 0.14$ μm, width $w_a = 1$ μm, and length $L_C = 260$ μm. (a) Temperature $T = 298$ K (25 °C), laser diode threshold current $I_{th} = 9.7$ mA, and $I_{fl} = 4$ mA. (b) Temperature $T = 328$ K (55 °C), laser diode threshold current $I_{th} = 21.5$ mA, and $I_{fl} = 10$ mA. Fluctuations into the lasing state enhance optical emission and remove carriers when $I_{inj} < I_{th}$ so that an additional current I_{fl} relative to the LED is required in the laser diode to reach the threshold carrier density, n_{th}. (Adapted with permission from [22]. Copyright 1993 IEEE.)

Bibliography

[1] For example, standard ISO 11554:2017 (en) https://www.iso.org/standard/69232.html

[2] For example, Levi A F J 2023 *Applied Quantum Mechanics* 3rd edn (Cambridge: Cambridge University Press)

[3] For reviews, see Walls D F 1983 *Nature* **306** 141
Loudon R and Knight P L 1987 *J. Mod. Opt.* **34** 709
Teich M C and Saleh B E A 1989 *Quantum Opt.* **1** 153
Davidovich L 1996 *Rev. Mod. Phys.* **68** 127
Scully M O and Zubairy M S 1997 *Quantum Optics* (Cambridge: Cambridge University Press)

[4] Nguyen L, Sloan J, Rivera N and Soljačić M 2023 *Phys. Rev. Lett.* **131** 173801

[5] Zhao S, Ding S, Huang H, Zaquine I, Fabre N, Belabas N and Grillot F 2024 *Phys. Rev. Res.* **6** L032021

[6] For example, Andersen U L, Gehring T, Marquardt C and Leuchs G 2016 *Phys. Scr.* **91** 053001

[7] For a somewhat more complete model, see Ahmed Yamada M and Saito M 2001 *IEEE J. Quantum Electron.* **37** 1600

[8] For example, Arakawa Y and Yariv A 1985 *IEEE J. Quantum Electron.* **21** 1666

[9] For example, Newell T C, Bossert D J, Stintz A, Fuchs B, Malloy K J and Lester L F 1999 *IEEE Photon. Technol. Lett.* **11** 1527

[10] Schawlow A L and Townes C H 1958 *Phys. Rev.* **112** 1940

[11] Lax M 1967 *Phys. Rev.* **160** 290

[12] Hempstead R D and Lax M 1967 *Phys. Rev.* **161** 350

[13] Henry C H 1982 *IEEE J. Quantum Electron.* **18** 259
Henry C H 1983 *IEEE J. Quantum Electron.* **19** 1391

[14] For example, Peterman K 1988 *Laser Diode Modulation and Noise* (Dordrecht: Kluwer)
Agrawal G P and Dutta N K 1993 *Semiconductor Lasers* (Dordrecht: Van Norstrand Reinhold)

[15] DeGiorgio V and Scully M O 1970 *Phys. Rev. A* **2** 1170

[16] Graham R and Haken H 1970 *Z. Phys.* **237** 31

[17] Grossmann S and Richter P H 1971 *Z. Phys.* **242** 458

[18] Corti M and Degiorgio V 1976 *Phys. Rev. Lett.* **36** 1173

[19] Pakhalov V B and Chirkin A S 1977 *Sov. J. Quantum Electron.* **7** 715

[20] Salomaa R and Stenholm S 1977 *Appl. Phys.* **14** 355

[21] O'Gorman J, Levi A F J, Schmitt-Rink S, Tanbun-Ek T, Coblentz D L and Logan R A 1992 *Appl. Phys. Lett.* **60** 157

[22] Chuang S L, O'Gorman J and Levi A F J 1993 *IEEE J. Quantum Electron.* **29** 1631

[23] O'Gorman J, Chuang S L and Levi A F J 1993 *Appl. Phys. Lett.* **62** 1454

IOP Publishing

Essential Semiconductor Laser Device Physics (Second Edition)

A F J Levi

Chapter 6

Quantized particle number

This chapter describes the influence photon and electron number quantization has on the behavior of small lasers. Photon particle number states interacting with a Fabry–Perot resonator and a multilayer dielectric filter are introduced along with some of the elements controlling single-photon dynamics in a Fabry–Perot resonator. The development of a semiclassical master equation with continuum probability functions to describe quantized particle number states in a mesolaser is demonstrated. The key parts of the semiclassical master equations used to predict the transient response and noise in a single-mode mesolaser are provided.

Classical concepts and theories [1] cannot explain some aspects of semiconductor device physics. When observed on a small enough time or length scale, an electron point *particle* of rest mass m_0, quantized spin magnitude $\hbar/2$, and discrete value of charge $-e$, can sometimes behave as if it is a *wave*. Similarly, a light *wave* of radial frequency ω can sometimes behave as if it is a *particle*. The *wave–particle duality* of an electron, the wave–particle duality of a photon, and other mysterious effects, such as the existence of *identical indistinguishable particles*, *linear superposition of particle states*, and quantum entanglement of particles, cannot be explained using classical Newtonian mechanics or Maxwell's equations. Fortunately, quantum mechanics provides a methodology for describing and predicting behavior. In contrast to the classical case, quantum system dynamics involve time evolution of quantum field amplitude and phase that cannot be measured directly.

In 1905 Einstein explained the photoelectric effect by postulating that light (in agreement with Planck's work) is quantized into particles, each of *energy $E = \hbar\omega$*. The quantum of light is a *photon*. A photon has zero mass[1], is *relativistic* because it travels at the speed of light, has quantized spin magnitude $\hbar$, obeys Bose statistics, and is an example of an elemental particle in quantum mechanics.

---

[1] Photon mass is the subject of ongoing research. If a photon has mass then its value is very small.

doi:10.1088/978-0-7503-6417-1ch6

A plane wave linearly polarized photon of wavelength λ_{ph} and magnitude of wave vector $|\mathbf{k}| = 2\pi/\lambda_{ph}$ has quantized energy $E = \hbar\omega$ and quantized momentum $\mathbf{p} = \hbar\mathbf{k}$. The wavelength in nm of a photon of energy $E\,(\mathrm{eV}) = \hbar\omega$ in free space is approximately

$$\lambda_{ph}(\mathrm{nm}) = \frac{1240}{E(\mathrm{eV})}. \tag{6.1}$$

6.1 An experiment to prove the photon exists

In 1977, many years after the discovery of quantum mechanics, the results of the first laboratory experiments proving the existence of the photon were published. Kimble, Dagenais, and Mandel showed that light is made up of discrete photons, each of which can create a single 'click' in detector output [2]. Later, in 1986, Grangier, Roger, and Aspect published the results of improved experiments as well as demonstrating the interference of a single photon with itself and, in this way, experimentally demonstrated the uniquely quantum mechanical particle and wave nature of the photon [3].

As illustrated in figure 6.1, the particle nature of the photon may be shown experimentally by using a laser diode-based single-photon source, a fiber-optic connected beam splitter, and two single-photon detectors. In its simplest form, attenuated emission from a laser diode with Poisson photon statistics can be used to approximate a single photon source. In this case, if the Poisson distribution has a mean of 0.1 then the probability of zero photons is 0.9, the probability of one photon is 0.09, and the probability of more than one photon is 0.01. Single-photon emission is spaced apart in time on average by $\langle \tau_{ph} \rangle$ and this photon stream is directed to an input port of an ideal lossless symmetric 50:50 beam splitter with a linear response. Each of the two output ports of the beam splitter is connected to single-photon

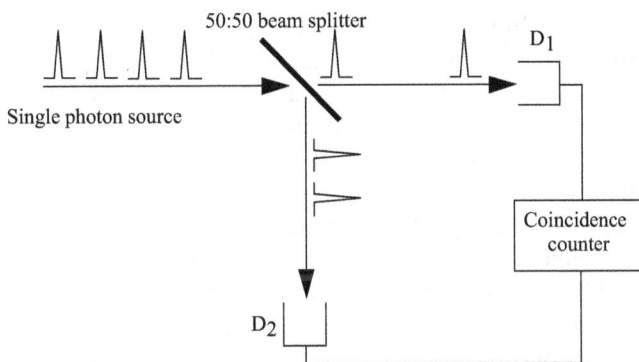

Figure 6.1. Schematic of the experimental arrangement to prove the photon exists. A stream of single photons is assumed to be present and incident on an ideal lossless symmetric 50:50 beam splitter. Single-photon detectors are placed at each output port of the beam splitter. If a single photon is an indivisible elementary particle then the detectors D_1 and D_2 record no coincidence counts because each photon must either be detected by D_1 or D_2, but not both.

detectors D_1 and D_2, respectively. Typically, each single-photon detector creates an output if one *or more* photons are present in a given measurement time interval. More sophisticated *photon number-resolving* detectors can provide an output proportional to the number of photons present and, in this way, multi-photon events can be eliminated. In the simplest configuration, the photon time of flight between each beam splitter output port and the corresponding detector is the same and the measurement time interval of each detector is much smaller than $\langle \tau_{ph} \rangle$. It is important to understand that the path taken by each indivisible photon in the system can only be *inferred after detection* by D_1 or D_2. This is a direct consequence of the standard Copenhagen interpretation of quantum mechanics in which the properties of any system are obtained *after* interaction of the quantum system with the measurement device (in this case, the measurement device consists of the detectors D_1 and D_2). In quantum mechanics, because there are two possible photon paths through the system, the selection of the inferred photon path is purely random and, hence, noncausal. In this experimental configuration, an indivisible quantized photon is either detected by D_1 or D_2, but not both, and so there are no coincidence counts recorded by the two detectors. It follows that the photon may be viewed as an indivisible, detectible, quantized particle.

Even if the flux of photon particles incident on the ideal lossless symmetric 50:50 beam splitter illustrated in figure 6.1 is noise-free so that individual photons are precisely spaced apart in time by a fixed τ_{ph}, there is a 50% chance that the photon will not be transmitted. This introduces quantum partition noise in the transmitted and reflected photon flux.

A convenient quantum mechanical model of the process at the beam splitter involves considering scattering from an input photon mode to an output mode. The boson annihilation operator $\hat{b}$ is applied to the photon state in the input mode, followed by the application of the creation operator $\hat{b}^\dagger$ to the vacuum state of the appropriate output mode.

The important result is that no single-photon coincidence counts are detected in the experiment. This cannot be explained by a classical wave description of electromagnetism given by Maxwell's equations. Maxwell's equations predict that electromagnetic waves introduced at one input port of a beam splitter appear at both output detectors simultaneously, resulting in coincidence detection, which directly conflicts with the experimental results. It is, of course, possible to connect the quantized photon description to the continuum classical description of electro-magnetism. This most naturally occurs when many incoherent photons contribute to a particular electromagnetic field. It is interesting to explore what is called the quantum–classical boundary, in which the transition between the most non-classical behavior and behavior that seems closest to classical expectations is studied. When there are very few photons present, special circumstances involving identical indistinguishable photons, or a coherent superposition of photons, then a quantum description is often most appropriate.

See appendix D for more details about what happens when one or more discrete identical, indistinguishable, photons are present at the input ports of a beam splitter.

6.2 The Fabry–Perot resonator

Figure 6.2 is a schematic that represents a lossless Fabry–Perot resonator consisting of two identical parallel mirrors separated by a length L_C. Each mirror has the same photon transmission and reflection coefficients and may be treated as a special case of a beam splitter. The complex transmission amplitude through the complete resonator system is related to photon reflection coefficient r_{ph} and phase $\phi = kL_C$ for light of wave number $k_{ph} = 2\pi/\lambda_{ph}$ and wavelength λ_{ph} via

$$T(r_{ph}, \phi) = \frac{(1 - |r_{ph}|^2)e^{-2i\sqrt{1-|r_{ph}|^2}}}{|r_{ph}|^2 e^{-2i\sqrt{1-|r_{ph}|^2}} e^{-2i\phi} - 1} \tag{6.2}$$

and the complex reflectance amplitude of the complete resonator is

$$R(r_{ph}, \phi) = \frac{|r_{ph}|e^{-i\sqrt{1-|r_{ph}|^2}}\left(e^{-i\phi} - e^{i\phi}e^{-2i\sqrt{1-|r_{ph}|^2}}\right)}{|r_{ph}|^2 e^{-2i\sqrt{1-|r_{ph}|^2}} e^{-2i\phi} - 1}. \tag{6.3}$$

The unitary resonator system requires $|T|^2 + |R|^2 = 1$ and $TR^* + RT^* = 0$. The fact that $|T|^2 + |R|^2 = 1$ is an expression of energy conservation in the lossless system.

The beam splitter formalism described in appendix D can be used to connect the amplitude of input and output states by a 2×2 matrix:

$$\begin{bmatrix} a_3 \\ a_4 \end{bmatrix}_{out} = \begin{bmatrix} T & R \\ R & T \end{bmatrix} \begin{bmatrix} a_1 \\ a_2 \end{bmatrix}_{in}. \tag{6.4}$$

A photon can be introduced at input port 1 using the operator $\hat{b}_1^\dagger$ to create state $|1\rangle_1 = \hat{b}_1^\dagger |0\rangle_1$ from the vacuum state $|0\rangle_1$. A single-mode n_{tot}-photon Fock (photon number) state at input port 1 is created by multiple applications of the creation operator to give $|n_{tot}\rangle_1 = \frac{(\hat{b}_1^\dagger)^{n_{tot}}|0\rangle_1}{\sqrt{n_{tot}!}}$. The vacuum state at port 2 is $|0\rangle_2$. Equation (6.4) shows that the Fabry–Perot resonator transforms this n_{tot}-photon input state

$$|n_1 = n_{tot}, \; n_2 = 0\rangle_{1,2} = \frac{\left(\hat{b}_1^\dagger\right)^{n_{tot}}}{\sqrt{n_{tot}!}} |0, \; 0\rangle_{1,2} \tag{6.5}$$

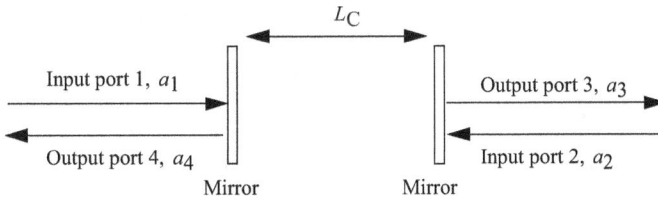

Figure 6.2. Diagram of two symmetric lossless mirrors configured as a Fabry–Perot resonator. Input and output ports are indicated. The resonator cavity has length L_C.

to the output state

$$|n_3, n_4\rangle_{3,4} = \frac{\left(T\hat{b}_3^\dagger + R\hat{b}_4^\dagger\right)^{n_{tot}}}{\sqrt{n_{tot}!}} |0, 0\rangle_{3,4}$$

$$= \frac{1}{\sqrt{n_{tot}!}} \sum_{l=0}^{n_{tot}} \binom{n_{tot}}{l} T^l R^{n-l} \sqrt{l!(n_{tot} - l)!} |l, n_{tot} - l\rangle_{3,4},$$

(6.6)

where $n_{tot} = n_1 + n_2$.

The probability of detecting j photons transmitted through the Fabry–Perot resonator to output port 3 is [4]

$$P_3(j, n_{tot}) = \frac{n_{tot}!}{j!(n_{tot} - j)!} |T|^{2j} (1 - |T|^2)^{n_{tot} - j}.$$

(6.7)

Because a multi-particle quantum state introduces an additional phase in a resonator, the photon number resolved transmission spectrum is non-classical. As illustrated in figure 6.3, the spectral linewidth of resonant n_{tot}-photon transmission is reduced compared to the classical case.

As with a beam splitter, transmission of a Fock state optical pulse will create photon number-dependent quantum partition noise.

Formal optimization methods may be used to find configurations of dielectrics that scatter classical electromagnetic radiation in a desired way. Such designs are often nonintuitive, have an aperiodic physical structure, and are robust to small variations in manufacture [5]. It is possible to use a similar approach to optimally

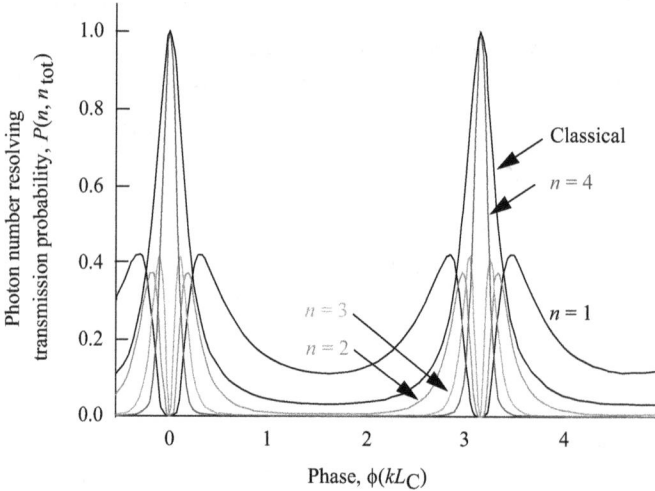

Figure 6.3. Calculated photon number resolved transmission spectrum for a Fabry–Perot resonator compared to classical light. The parameters are the total number of incident photons in the Fock state $n_{tot} = 4$, mirror reflectivity $|r_{ph}| = 0.837$, cavity length $L_C = 5\lambda_0$, and resonant wavelength $\lambda_0 = 1.5\,\mu m$.

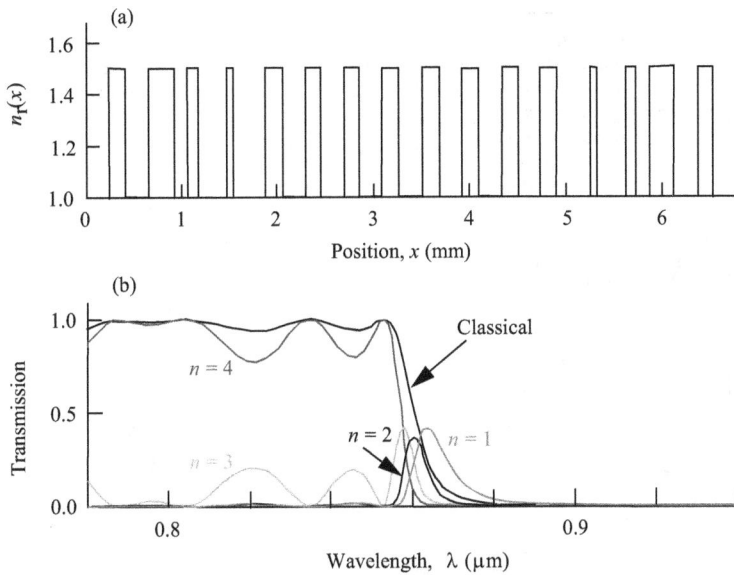

Figure 6.4. (a) Dielectric layer structure for an optical filter and (b) corresponding calculated photon number resolved spectra compared to illumination with classical light. The parameters are the total number of incident photons in a single-mode Fock state $n_{tot} = 4$, refractive index $n_{r,\,1} = 1$ and $n_{r,\,1} = 1.5$, and 16 dielectric layer pairs arranged symmetrically about the mid-point. The fact that photon number resolved spectral features can be sharper than classical light allows more information to be encoded in dielectric structures probed by quantum light than classical light in the presence of a given background noise level.

design dielectric structures to have maximally non-classical photon number resolved spectra. Figure 6.4(a) is an example of a multi-layered dielectric optical filter, and figure 6.4(b) shows the corresponding calculated photon number resolved transmission spectra compared to illumination with classical light. It is apparent that more readable information can be encoded in dielectric structures probed by quantum light than classical light in the presence of a given background noise level.

6.3 Control of single-photon dynamics in a Fabry–Perot resonator

The equations describing the behavior of a single-photon wave function $\psi(x, t)$ are the same as those for a classical electromagnetic field. However, rather than interpreting the magnitude squared of a single-photon wave function as the probability of finding a photon *particle* at position x, the value of $|\psi(x, t)|^2$ may be used to describe single-photon *energy* probability density at position x [6]. Because the real and imaginary parts of the single-photon wave function $\psi(x, t)$ obey Maxwell's equations, methods used to solve classical electromagnetic problems may be applied to the case of a single photon.

Control of photon dynamics in a classical or quantum system may be achieved in different ways. Figure 6.5(a) illustrates an example of *closed-loop* control applied to a quantum system. Figure 6.5(b) shows how ray tracing can be used to efficiently

(a)

```
        ┌─────────────────────┐
        │ Physical objective and │
        │ initial parameter values │
        └─────────────────────┘
                    │
                    ▼
        ┌─────────────────────┐
        │ Model-based control   │
        │ parameters and        │
        │ control-field generator │
        └─────────────────────┘
                    │
┌──────────────┐    ▼
│Change control │  ┌─────────────────────┐
│parameters     │  │ Quantum system with  │
└──────────────┘  │ controlled dynamics  │
                   └─────────────────────┘
                            │
                            ▼
                  ┌─────────────────────┐
                  │      Detector        │
                  └─────────────────────┘
                            │
                            ▼
                       ◇ Objective
              No        met? ◇
                            │
                          Yes ▼
                  ┌─────────────────────┐
                  │        End           │
                  └─────────────────────┘
```

(b)

Graphical ray-tracing to control energy density in resonator

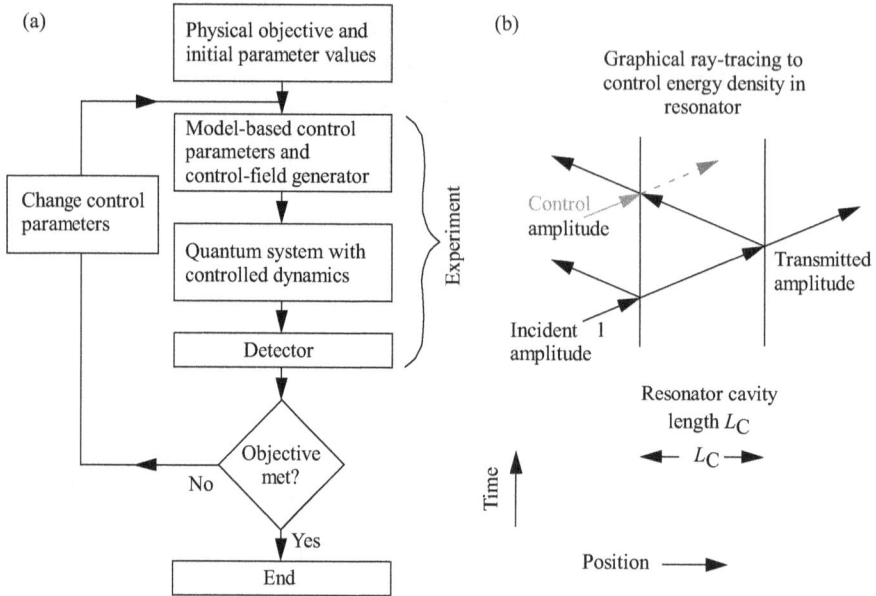

Control amplitude

Transmitted amplitude

Incident 1 amplitude

Resonator cavity length L_C

← L_C →

Time

Position ⟶

Experiment

Figure 6.5. (a) Illustration of closed-loop optimization of system dynamics. (b) Efficient open-loop optimization of a coherent classical or single-photon pulse interacting with a Fabry–Perot resonator. A graphical method can be used to find the near-optimal sequence of pulses to control the interaction.

find parameters for *open-loop* control of coherent photons interacting with a Fabry–Perot resonator.

For the case of a coherent classical or single-photon pulse interacting with a Fabry–Perot resonator a sequence of pulses can be used to control the interaction. As shown in appendix E, formal optimization techniques are not initially required. They can be replaced by a simple intuitive graphical method (in this case, ray tracing) to efficiently find near-optimal control pulse sequences to meet a desired objective. Formal methods can then be used to fine-tune the control pulse shape for optimal control.

It is possible to use photon pulse sequences interacting with a Fabry–Perot resonator to store and release photon energy from the resonator in a precise and deterministic way, and it is also possible to perform other useful information processing functions [7]. However, while elegant, this does not directly contribute to a model capable of describing the more complicated behavior of a mesoscale laser. An initial step in this direction is considered next.

6.4 Particle number quantization in a mesolaser

The description of mesolaser behavior can be made more accurate by incorporating the contribution of fluctuations and correlations due to the quantized excitation of electronic states and photon numbers. The quantized particle number states can be modeled by probability functions that are part of a semiclassical master equation in which energy and particle number conservation are explicitly taken into account [8, 9].

However, the quantized particle energy states in the master equation contain no phase information, so only the steady-state and transient particle number contribution to quantum fluctuations can be studied [10, 11]. It is found that the presence of such fluctuations in small devices can result in bimodal and non-Poisson probability distributions for both excited states and photon numbers. The same fluctuations in mesolasers act to suppress lasing and enhance spontaneous emission near the threshold. Because the semiclassical master equations contain no information on phase fluctuations, it is not possible to describe other subtle correlation effects, such as sub-threshold symmetry-protected entangled quantum states.

The photon states described in the previous sections of this chapter contain particle amplitude *and* phase information. Interaction of photons with a dielectric is treated classically using a lossless refractive index. Because a full quantum model of photons interacting with atoms in a small semiconductor laser diode is numerically very intensive, it makes sense to explore simpler semiclassical models initially.

Very small (mesoscale) laser diode behavior can be approximated using a semiclassical master equation that quantizes photon and excitation particle numbers but does *not* make use of phase information. In this approach, the evolution of discrete carrier and photon particle numbers is determined by evaluating continuum probability functions describing the integer number of excited electronic states (carriers), n, and the integer number of photons, s, in the lasing mode of frequency ω. It is assumed that the state (n, s) can describe the system.

As shown in figure 6.6, transitions to and from the state (n, s) can change the probability distribution, $P_{n,s}$. A semiclassical master equation that describes the time evolution of the probability $P_{n,s}$ associated with state (n, s) is

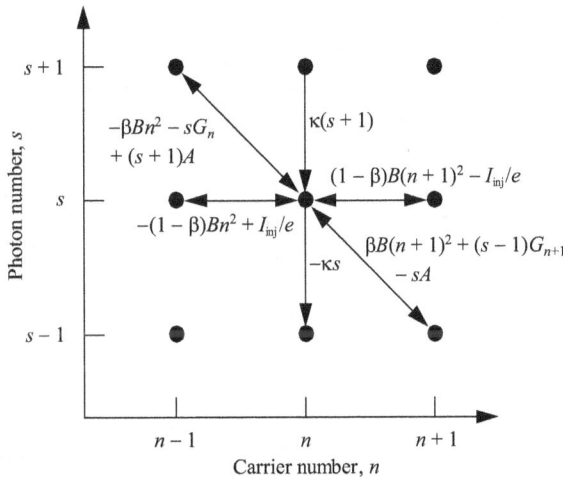

Figure 6.6. Diagram illustrating transition rates in and out of number quantized state (n, s). A positive sign indicates flow into the state and a negative sign indicates flow out of the state. The parameters are described in the text.

$$\dot{P}_{n,s} \equiv \frac{dP_{n,s}}{dt} = - \kappa(sP_{n,s} - (s+1)P_{n,s+1}) - (sG_n P_{n,s} - (s-1)G_{n+1}P_{n+1,s-1})$$

$$- (sAP_{n,s} - (s+1)AP_{n-1,s+1}) - \beta B(n^2 P_{n,s} - (n+1)^2 P_{n+1,s-1}) \quad (6.8)$$

$$- (1-\beta)B(n^2 P_{n,s} - (n+1)^2 P_{n+1,s}) - \frac{I_{\text{inj}}}{e}(P_{n,s} - P_{n-1,s})$$

and may be solved by integration. In this expression, a positive sign indicates flow into the state (n, s) and a negative sign indicates flow out of the state. B is the spontaneous emission coefficient,

$$sG_n = \frac{G_{\text{slope}} \Gamma_{\text{opt}} \, cns}{n_r V_{\text{vol}}} \quad (6.9)$$

is the stimulated emission rate in the system at photon energy $\hbar\omega$,

$$sA = \frac{G_{\text{slope}} \Gamma_{\text{opt}} \, cn_{\text{ot}}s}{n_r V_{\text{vol}}} \quad (6.10)$$

is the stimulated absorption rate, n_{ot} is the transparency carrier number, c is the speed of light in a vacuum, Γ_{opt} is the overlap of the optical field intensity with the gain medium, G_{slope} is the gain slope coefficient, β is the fraction of spontaneous emission into the lasing mode, and n_r is the refractive index of the active volume V_{vol}. If the mesolaser uses a Fabry–Perot cavity then the total optical loss rate is

$$\kappa s = \frac{c}{n_r}\left(\alpha_i + \frac{1}{2L_C}\ln\left(\frac{1}{r_1 r_2}\right)\right)s, \quad (6.11)$$

where $r_{1,2}$ is the mirror reflectivity, α_i is the internal loss and L_C is the cavity length. The term $-\beta Bn^2 P_{n,s}$ describes spontaneous emission of photons involving transitions from state (n, s) to state $(n-1, s+1)$ where B is the spontaneous emission coefficient. The term $-sG_n P_{n,s}$ describes the stimulated emission of photons from the state (n, s) to the state $(n-1, s+1)$ and G_n is the stimulated emission coefficient. $-(1-\beta)Bn^2 P_{n,s}$ is the decay of electrons into nonlasing photons via transitions from state (n, s) to state $(n-1, s)$. The term $I_{\text{inj}}P_{n,s}$ describes the pumping of electrons into the system, causing transitions from state (n, s) to state $(n+1, s)$. $I_{\text{inj}} = I_{\text{inj}}(t)$ is the time-dependent diode injection current. Stimulated absorption of photons involving transitions from state (n, s) to state $(n+1, s-1)$ is represented by $-sAP_{n,s}$ where A is the stimulated absorption coefficient. The term $-\kappa sP_{n,s}$ describes the decay of cavity photons in which transitions from state (n, s) to state $(n, s-1)$ occur, where κ is the optical loss coefficient.

The average values of carrier and photon numbers evolve similarly to the predictions of the continuum mean-field single-mode rate equations. However, the higher-order moments in the distribution of $P_{n,s}$ during transients and at low pump levels deviate substantially from normal behavior. The equations for average behavior can be derived from the semiclassical master equation by multiplying by the number of excited electronic states, n, and the number of photons, s, and averaging. For $\langle n \rangle$ this gives

$$\frac{d}{dt}\langle n\rangle = \frac{I_{inj}}{e} - B\langle n^2\rangle - \frac{\Gamma_{opt}G_{slope}c}{V_{vol}n_r}\langle(n - n_{ot})s\rangle \tag{6.12}$$

and for $\langle s\rangle$

$$\frac{d}{dt}\langle s\rangle = \frac{\Gamma_{opt}G_{slope}c}{V_{vol}n_r}\langle(n - n_{ot})s\rangle - \kappa\langle s\rangle + \beta B\langle n^2\rangle. \tag{6.13}$$

Continuum mean-field single-mode rate equations are recovered *if* the correlations $\langle(n - n_{ot})s\rangle$ and $\langle n^2\rangle$ *factorize* so that $\langle(n - n_{ot})s\rangle = \langle(n - n_{ot})\rangle\langle s\rangle$ and $\langle n^2\rangle = \langle n\rangle^2$. However, in a single-mode mesolaser diode, the dynamic response of n and s is strongly correlated so that $\langle(n - n_{ot})s\rangle$ and $\langle n^2\rangle$ cannot be factorized.

Consider solving the semiclassical master equation for a step change in the injection current I_{inj} starting from zero at time $t < 0$ with zero electrons $n = 0$ and zero photons $s = 0$ in the device. The initial condition is such that the probability is $P_{n=0, s=0}(t < 0) = 1$. A constant current I_{inj} turned on when $t \geqslant 0$ has the effect of introducing an electron at a fixed average rate, $1/t_e$. For simplicity, precisely one electron is introduced (noise-free) at exactly every time interval t_e with the first electron introduced at time $t = 0$. Whether the electrons are introduced noise-free or stochastically makes little difference for the parameters of practical importance. After the first electron is introduced at time $t = 0$ the probabilities for possible states of the system at time $t = t_e$ are $P_{n=0, s=0}$, $P_{n=0, s=1}$, and $P_{n=1, s=0}$. The maximum number of particles in the system is one. The probability of $P_{n=0, s=0}$ is finite at time $t = t_e$ because the system is open so an electron can be converted to a photon and the photon lost from the cavity. The master equation for probability $P_{n=0, s=0}$ is

$$\dot{P}_{0,0} = -\kappa((0)P_{0,0} - ((0) + 1)P_{0,1}) - ((0)G_0P_{0,0} - ((0) - 1)G_1P_{1,-1})$$

$$-((0)AP_{0,0} - ((0) + 1)AP_{-1,1}) - \beta B((0)^2P_{0,0} - (1)^2P_{1,-1})$$

$$-(1 - \beta)B((0)^2P_{0,0} - ((0) + 1)^2P_{1,0}) - \frac{I_{inj}}{e}(P_{0,0} - P_{-1,0}). \tag{6.14}$$

Terms with negative values of n or s are not allowed and so set to zero. The equation becomes

$$\dot{P}_{0,0} = \kappa(1)P_{0,1} + (1 - \beta)B(1)^2P_{1,0} - \frac{I_{inj}}{e}P_{0,0}. \tag{6.15}$$

Similarly, after finding the equations for $P_{n=0, s=1}$, and $P_{n=1, s=0}$, the system of first-order coupled differential equations can be written as a matrix

$$\begin{bmatrix} \dot{P}_{0,0} \\ \dot{P}_{0,1} \\ \dot{P}_{1,0} \end{bmatrix} = \begin{bmatrix} -\dfrac{I_{inj}}{e} & \kappa(1) & (1 - \beta)B(1)^2 \\ 0 & -\kappa(1) - A(1) & \beta B(1)^2 \\ \dfrac{I_{inj}}{e} & A & -B(1)^2 \end{bmatrix} \begin{bmatrix} P_{0,0} \\ P_{0,1} \\ P_{1,0} \end{bmatrix} \tag{6.16}$$

and integrated over the first time interval using the fourth-order Runge–Kutta method.

At the end of the next time interval the possible states of the system are $P_{n=0,\,s=0}$, $P_{n=0,\,s=1}$, $P_{n=1,\,s=0}$, $P_{n=0,\,s=2}$, $P_{n=1,\,s=1}$, and $P_{n=2,\,s=0}$. The maximum number of particles in the system is two. Terms $P_{n=0,\,s=2}$, $P_{n=1,\,s=1}$, and $P_{n=2,\,s=0}$ have zero value when they are introduced and subsequently evolve to finite values by coupling to other matrix elements. The system of first-order coupled differential equations is

$$
\begin{bmatrix} \dot{P}_{0,0} \\ \dot{P}_{0,1} \\ \dot{P}_{1,0} \\ \dot{P}_{0,2} \\ \dot{P}_{1,1} \\ \dot{P}_{2,0} \end{bmatrix} =
\begin{bmatrix}
-\dfrac{I_{\text{inj}}}{e} & \kappa & (1-\beta)B & 0 & 0 & 0 \\
0 & -\kappa - A - \dfrac{I_{\text{inj}}}{e} & \beta B & 2\kappa & (1-\beta)B & 0 \\
\dfrac{I_{\text{inj}}}{e} & A & -B - \dfrac{I_{\text{inj}}}{e} & 0 & \kappa & 4(1-\beta)B \\
0 & 0 & 0 & -2(\kappa - A) & G_1 + \beta B & 0 \\
0 & \dfrac{I_{\text{inj}}}{e} & 0 & 2A & -\kappa - G_1 - B & 4(1-\beta)B \\
0 & 0 & \dfrac{I_{\text{inj}}}{e} & 0 & A & 4(1-\beta)B
\end{bmatrix}
\begin{bmatrix} P_{0,0} \\ P_{0,1} \\ P_{1,0} \\ P_{0,2} \\ P_{1,1} \\ P_{2,0} \end{bmatrix}. \quad (6.17)
$$

The system of first-order coupled differential equations at the Nth step is

$$
\begin{bmatrix} \dot{P}_{0,0} \\ \dot{P}_{0,1} \\ \dot{P}_{1,0} \\ \dot{P}_{0,2} \\ \dot{P}_{1,1} \\ \dot{P}_{2,0} \\ \vdots \\ \dot{P}_{N-1,1} \\ \dot{P}_{N,0} \end{bmatrix} =
\begin{bmatrix}
-\dfrac{I_{\text{inj}}}{e} & \kappa & (1-\beta)B & 0 & \cdots & 0 & 0 \\
0 & -\kappa - A - \dfrac{I_{\text{inj}}}{e} & \beta B & 2\kappa & \cdots & 0 & 0 \\
\dfrac{I_{\text{inj}}}{e} & A & -B - \dfrac{I_{\text{inj}}}{e} & 0 & \cdots & 0 & 0 \\
0 & 0 & 0 & -2(\kappa - A) & \cdots & 0 & 0 \\
\vdots & \vdots & \vdots & \vdots & \vdots \vdots & \vdots & \vdots \\
0 & 0 & 0 & 0 & \cdots & \vdots & (1-\beta)BN^2 \\
0 & 0 & 0 & 0 & \cdots & A & -BN^2
\end{bmatrix}
\begin{bmatrix} P_{0,0} \\ P_{0,1} \\ P_{1,0} \\ P_{0,2} \\ P_{1,1} \\ P_{2,0} \\ \vdots \\ P_{N-1,1} \\ P_{N,0} \end{bmatrix}. \quad (6.18)
$$

Total probability is conserved at every step, and the states allowed are those of continuous-time Markov chains.

After addition of N electrons, the matrix has dimension $(N+1)^2$ and so the matrix grows quadratically with the number of injected electrons. The number of zero entries in the $(N+1)^2$ matrix is

$$
\frac{(N+1)^2 - (N+1)}{2} = \frac{(N^2 + N)}{2}. \quad (6.19)
$$

It follows that the number of non-zero entries in the matrix is

$$
(N+1)^2 - \frac{(N^2+N)}{2} = \frac{N^2}{2} + \frac{3N}{2} + 1 = N\left(\frac{N}{2} + \frac{3}{2}\right) + 1. \quad (6.20)
$$

Figure 6.7 shows the calculated probability $P_{n,s}$ (using a log-scale) in response to a step change in injection current from $I_{\text{inj}}(t \leqslant 0)$ to $I_{\text{inj}}(t > 0)$ at time (a) $t = 0.4$ ns, (b) $t = 0.8$ ns, (c) $t = 1.2$ ns, and (d) $t = 1.6$ ns. The bimodal nature of the photon distribution is apparent with a large peak for zero photon probability both during

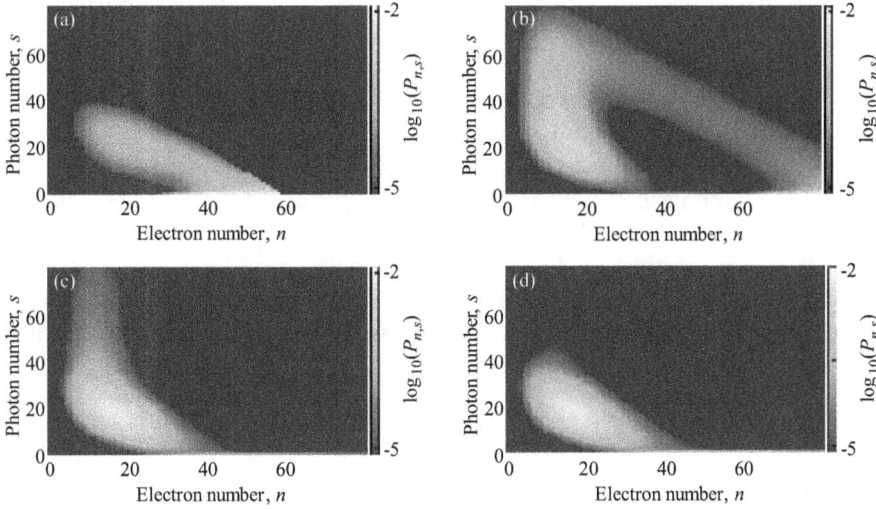

Figure 6.7. Calculated transient $P_{n,s}$ in response to a step change in injection current from $I_{\text{inj}} = 0$ nA to $I_{\text{inj}} = 24$ nA at time (a) $t = 0.4$ ns, (b) $t = 0.8$ ns, (c) $t = 1.2$ ns, and (d) $t = 1.6$ ns. The color scale is $\log_{10}(P_{n,s})$. Parameters are: $r_1 = 1$, $r_2 = 0.998$, $L_C = 10^{-5}$ cm, $V_{\text{vol}} = 10^{-17}$ cm^{-3}, $n_r = 4$, $\Gamma_{\text{opt}} = 0.25$, $A_{\text{nr}} = 0$, $B \times V_{\text{vol}} = 10^{-10}$ cm^3 s^{-1}, $C = 0$, $\beta = 0.1$, $G_{\text{slope}} = 7.5 \times 10^{-16}$ cm^2 s^{-1}, $I_{\text{inj}} = 24$ nA, $n_{\text{ot}} = 10^{18}$ cm^{-3}.

and after the transient associated with the device response to the step change in injection current, I_{inj}.

The system evolves by transitioning between neighboring states via the processes indicated in figure 6.6. A Monte-Carlo trajectory can be used to create a time-based evolution of the system. In this case, the time constants, τ_i, of all possible independent transitions involving the state (n, s) are calculated, and the next time step is calculated using $t_i = -\tau_i \ln(\text{rand})$ where the subscript labels the channel and rand is a uniformly distributed random number between zero and one. The channel with the lowest t_i is chosen and the system makes a move to the new state in time t_i. The process involves a series of biased random transitions on a $n - s$ grid whose trajectories sample the continuous probability function $P_{n,s}$ for each state (n, s). The steady-state probability distribution for a particular injection current is obtained by averaging over multiple trajectories, where each trajectory consists of millions of time steps. The probability of state (n, s) is $P_{n,s}$. This probability is estimated from the relative time spent in state (n, s).

Figure 6.8 shows steady-state characteristics for lasers with different active volumes V_{vol} and spontaneous emission factors β. Continuum mean-field rate-equation results are compared with those obtained using the Monte-Carlo method.

Figure 6.8(a) shows the results for a conventional Fabry–Perot laser diode with spontaneous emission factor $\beta = 5 \times 10^{-5}$ and active volume $V_{\text{vol}} = 33.6$ μm^3. The expected classical laser threshold behavior and carrier (electronic excitation) pinning above the threshold are observed. The total optical output power in mW at an operating emission wavelength of 1310 nm can be determined by multiplying the photon number by 5.2×10^{-5}.

Figure 6.8(b) gives results for a microdisk laser with active volume $V_{vol} = 0.12 \, \mu m^3$. Optical output power in μW at a 1310 nm wavelength is obtained by multiplying the photon number by 7.1×10^{-3}. The steady-state characteristics show that the change in slope of photon expectation number $\langle s \rangle$ around the phase transition region is considerably smoothed due to the large value of the spontaneous emission factor, $\beta = 0.1$.

Figure 6.8(c) gives the steady-state characteristics of a laser where the active volume has been reduced to $V_{vol} = 10^{-4} \, \mu m^3$ and $\beta = 10^{-4}$. Suppression of lasing is observed along with depinning of carriers in this limit in which there is both a very small active volume and a small value of β. Optical output power in nW at a wavelength of 800 nm is obtained by multiplying the photon number by 0.186.

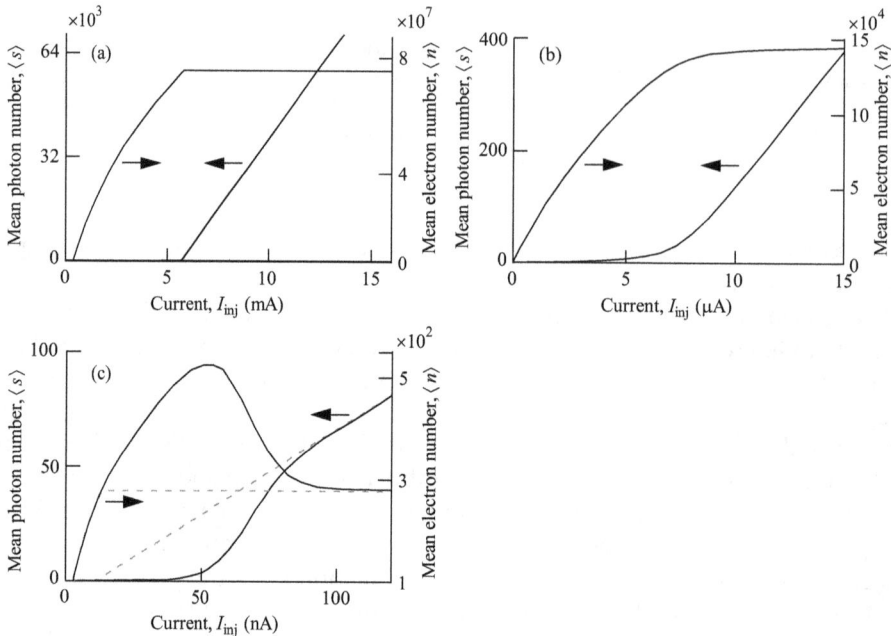

Figure 6.8. Calculated steady-state characteristics of macroscale and mesoscale semiconductor lasers showing the mean photon and electron number in the device plotted as a function of injection current. Results from a Monte-Carlo calculation (solid line) for (a) a Fabry–Perot laser, (b) a microdisk laser, (c) with mesoscale active volume. The dashed line is a comparison to the predictions of a mean-field rate-equation calculation. Parameters for (a): $V_{vol} = (300 \times 0.8 \times 0.14) \, \mu m^3$, $\Gamma_{opt} = 0.25$, $G_{slope} = 2.5 \times 10^{-16} \, cm^2 \, s^{-1}$, $A_{nr} = 2 \times 10^8 \, s^{-1}$, $B = 2.5 \times 10^{-10} \, cm^3 \, s^{-1}$, $C = 10^{-29} \, cm^6 \, s^{-1}$, $n_{ot} = 10^{18} \, cm^{-3}$, $\alpha_i = 40 \, cm^{-1}$, $n_r = 3.3$, $r_{1,2} = 0.32$, $\beta = 5 \times 10^{-5}$. Parameters for (b): $V_{vol} = (\pi \times (0.8)^2 \times 0.06) \, \mu m^3$, $\Gamma_{opt} = 0.25$, $G_{slope} = 2.5 \times 10^{-16} \, cm^2 \, s^{-1}$, $A_{nr} = 2 \times 10^8 \, s^{-1}$, $B = 10^{-10} \, cm^3 \, s^{-1}$, $C = 10^{-29} \, cm^6 \, s^{-1}$, $n_{ot} = 10^{18} \, cm^{-3}$, $\alpha_i = 10 \, cm^{-1}$, $n_r = 4$, $r_{1,2} = 0.999$, $\beta = 10^{-1}$. Parameters for (c): $V_{vol} = (0.1 \times 0.1 \times 0.01) \, \mu m^3$, $\Gamma_{opt} = 0.25$, $G_{slope} = 2.5 \times 10^{-18} \, cm^2 \, s^{-1}$, $A_{nr} = 2 \times 10^8 \, s^{-1}$, $B = 10^{-10} \, cm^3 \, s^{-1}$, $C = 10^{-29} \, cm^6 \, s^{-1}$, $n_{ot} = 10^{18} \, cm^{-3}$, $\alpha_i = 1 \, cm^{-1}$, $n_r = 4$, $r_{1,2} = 1 - 10^{-6}$, $\beta = 10^{-4}$.

Carrier (electronic excitation) depinning near the threshold may be investigated further by accounting for the spontaneous emission of photons into a nonlasing channel. The $(1 - \beta)$ term in figure 6.6 populates a nonlasing channel consisting of one or more optical modes that contain a total of s'' spontaneously emitted photons.

Figure 6.9 shows trajectories calculated in the time domain for different injection currents for a laser with a small active volume and a *very small* value of spontaneous emission factor, β. The $(1 - \beta)$ spontaneous emission terms are also included in the calculation but not shown in the figure. Figures 6.9(a) and (b) predict short bursts of photons (photon 'blinking') because of the presence of large quantum fluctuations. The value of the injection current is enough to cause fluctuations in the lasing state but not enough to support continuous lasing. With increasing injection current, the photon bursts last for a longer time and contribute to a bimodal average electron distribution. This is illustrated by the data in figure 6.9(c) and resembles the telegraph noise found in electrical circuits. For operation near the laser threshold, the semiclassical master equation calculation shows that switching occurs between two different characteristic system states—a physical effect not captured by a conventional continuum mean-field calculation.

An explanation of lasing suppression, enhanced spontaneous emission, and the depinning of carriers (electronic excitations) near the lasing threshold in a mesolaser may be found by studying the results shown in figure 6.9(c). The non-equilibrium open system driven by a constant injection current cannot continuously lase in this operating condition and switches between the lasing and nonlasing state. A random

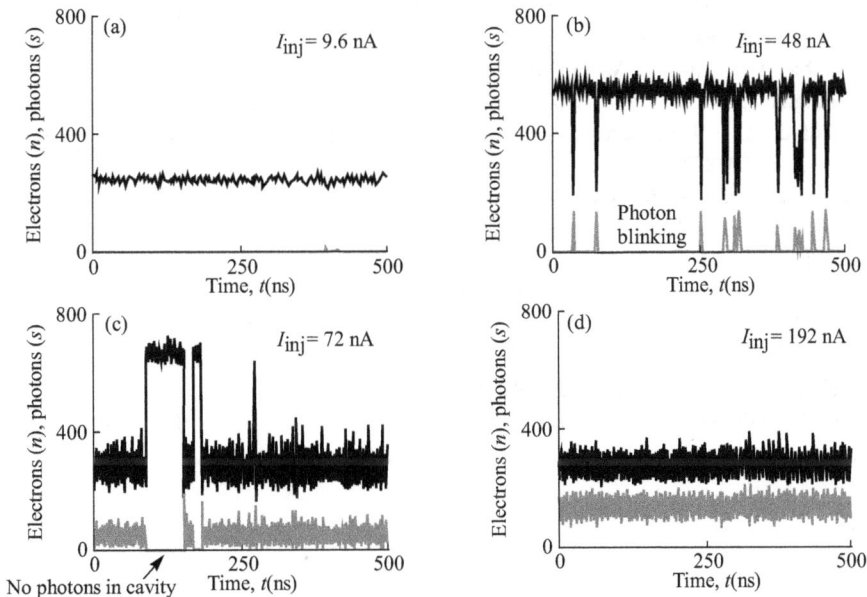

Figure 6.9. Time evolution of electrons (n, black) and photons (s, red) calculated by a random trajectory. (a) Injected current, $I_{inj} = 9.6$ nA, (b) $I_{inj} = 48$ nA, (c) $I_{inj} = 72$ nA, (d) $I_{inj} = 192$ nA. The parameters are as in figure 6.8(c).

spontaneous emission event is required to reinitiate lasing from the nonlasing to the lasing state. Lasing, being predominantly a stimulated process, requires the presence of photons in the lasing mode of the cavity. A larger active volume with a larger number of electrons has more spontaneous emission events, which prevents lasing shutdown. A smaller active volume, with a lesser likelihood of such events and a small value of β, experiences suppression of continuous lasing. Lasing suppression is accompanied by the depinning of carriers, which in turn results in enhanced spontaneous emission.

Figure 6.9(d) is an example of relatively strong lasing in a mesoscale device with quantized photon fluctuations about a mean value $\langle s \rangle = 136$. This behavior compares closely to the Langevin trajectories generated by adding Gaussian noise to the continuum mean-field rate equations. The average output power from both cavity mirrors is around 25 nW at an operating wavelength of 0.8 μm. In this case, the average number of photons in the nonlasing channel is $\langle s'' \rangle = 9.5$.

There is less noise in carrier number n in figure 6.9(a) because there are essentially only carriers (electronic excitations) in the system and very few lasing photons. As illustrated in figure 6.9(c), when both photons and electrons are in the system, the electron noise is enhanced because photon noise couples to the electron distribution. When the cavity empties of photons, the number of electrons increases, but the noise decreases. This is because noise coupled into the electron system from the photons is no longer present.

Bibliography

[1] Levi A F J 2016 *Essential Classical Mechanics for Device Physics* (San Rafael, CA: IOP/ Morgan Claypool)
[2] Kimble H J, Dagenais M and Mandel L 1977 *Phys. Rev. Lett.* **39** 691
[3] Grangier P, Roger G and Aspect A 1986 *Europhys. Lett.* **1** 173
[4] Wildfeuer C F, Pearlman A J, Chen J, Jingyun F, Migdall A and Dowling J P 2009 *Phys. Rev.* A **80** 043822
[5] Seliger P, Mahvash M, Wang C and Levi A F J 2006 *J. Appl. Phys.* **100** 034310
[6] Bialynicki-Birula I 1994 *Acta Phys. Pol.* **86** 97
   Smith B J and Raymer M G 2007 *New J. Phys.* **9** 414
   Raymer M G and Polakos P 2023 *Acta Phys. Pol.* A **143** S28
[7] Levi A F J, Campos Venuti L, Albash T and Haas S 2014 *Phys. Rev.* A **90** 022119
[8] Carmichael H J 1999 *Statistical Methods in Quantum Optics 1* (Berlin: Springer)
[9] Rice P R and Carmichael H J 1994 *Phys. Rev.* A **50** 4318
[10] Roy-Choudhury K, Haas S and Levi A F J 2009 *Phys. Rev. Lett.* **102** 053902
[11] Roy-Choudhury K and Levi A F J 2010. *Phys. Rev.* A **81** 013827

IOP Publishing

Essential Semiconductor Laser Device Physics (Second Edition)

A F J Levi

Chapter 7

Quantum behavior

In this chapter, essential elements of the quantization of the photon field and atom are introduced. The Jaynes–Cummings Hamiltonian and a two-level system in the rotating wave approximation are described. The key aspects of quantum behavior in single-mode mesolasers that include predictions of the steady-state mean photon number, photon Fano factor, photon emission linewidth, saturation in emitter inversion, long-lived sub-threshold dark states, and self-quenching as a function of pump rate and number of atoms are provided. The classical-to-quantum transition and the practical limitations of mesolasers are also discussed.

Describing a mesoscale laser in terms of quantized particle number states is a useful initial approach to understanding the role of fluctuations in determining device behavior. Semiclassical master equations, in which the time evolution of continuum probability functions describes the occupation of quantized particle excited electronic emitter number and photon number states, are a convenient framework in which to study the coupled electronic-photon system. It is found that quantum fluctuations in particle number dominate the steady-state and transient response in mesolasers. When the system is near the laser threshold, the semiclassical master equations predict non-Poisson probability distributions for excited electronic number states and photon number states. Experimentally accessible parameters can be found in which fluctuations near the laser threshold act to suppress lasing and enhance spontaneous emission.

However, there is more to the behavior of a mesoscale laser than the effects predicted by semiclassical master equations. A complete quantum description should be able to model quantized photon–atom interaction and include correlation and phase fluctuation effects. In very small semiconductor lasers with only a few atoms contributing to lasing, an external reservoir driving electronic excitation at high pump rates in the system is expected to induce self-quenching and thereby limit laser light output. When driven at pump rates below the self-quenching limit, long-lived emitter states emerge, the existence of which is revealed by peaks in photon

doi:10.1088/978-0-7503-6417-1ch7

fluctuations (the Fano-factor) as a function of pump rate. These and other effects are purely quantum and cannot be explained by semiclassical master equations.

7.1 Quantization of the photon field and atom

As described in chapter 6, quantum light interacting with a macroscopic resonant dielectric structure has unique non-classical behavior. This occurs even though the dielectric is modeled using a classical refractive index. Further exploration of light–matter interaction can benefit from a model in which *both* light in the form of photons *and* matter in the form of atoms are quantized. This is considered next.

In a *linear*, homogeneous (uniform in all positions), isotropic (uniform in all directions) medium of volume V_{vol}, an electromagnetic field can be expanded in a plane wave basis with wave vectors $\mathbf{k}$ and unit-normal polarization vector $\mathbf{e}_{\underset{\sim}{\mathbf{k}}}$. The Hamiltonian of the quantized electromagnetic (photon) field, H_{field}, can be described as the sum of quantized harmonically oscillating modes of frequency $\omega_{\mathbf{k}}$ so that

$$H_{\text{field}} = \sum_{\mathbf{k}} \hbar \omega_{\mathbf{k}} \left(\hat{b}_{\mathbf{k}}^{\dagger} \hat{b}_{\mathbf{k}} + \frac{1}{2} \right), \tag{7.1}$$

where $\hat{b}_{\mathbf{k}}^{\dagger}$ and $\hat{b}_{\mathbf{k}}$ are the harmonic oscillator boson creation and annihilation operators, respectively. These operators are dimensionless and satisfy the commutation relations $[\hat{b}_{\mathbf{k}}, \hat{b}_{\mathbf{k}}^{\dagger}] = 1$ and $[\hat{b}_{\mathbf{k}}, \hat{b}_{\mathbf{k}}] = [\hat{b}_{\mathbf{k}}^{\dagger}, \hat{b}_{\mathbf{k}}^{\dagger}] = 0$.

For an electromagnetic field in a vacuum with eigenfrequency $\omega_{\mathbf{k}}$ the quantized electric field operator is

$$\mathbf{E}(\mathbf{r}, \ t) = \sum_{\mathbf{k}} \mathbf{e}_{\underset{\sim}{\mathbf{k}}} \sqrt{\frac{\hbar \omega_{\mathbf{k}}}{2\varepsilon_0 V_{\text{vol}}}} \left(\hat{b}_{\mathbf{k}} e^{-i\omega_{\mathbf{k}}t + i\mathbf{k}\cdot\mathbf{r}} + \hat{b}_{\mathbf{k}}^{\dagger} e^{i\omega_{\mathbf{k}}t - i\mathbf{k}\cdot\mathbf{r}} \right). \tag{7.2}$$

7.1.1 The Jaynes–Cummings Hamiltonian

In a system consisting of a single atom interacting with a quantized electromagnetic field, the Hamiltonian is the sum of contributions from the atom, field, and interaction:

$$H = H_{\text{atom}} + H_{\text{field}} + H_{\text{int}}, \tag{7.3}$$

where H_{atom} is the Hamiltonian describing states of an electron in the isolated atom and H_{field} is the Hamiltonian describing the quantized radiation field. The eigenstates $|i\rangle$ of the isolated atom satisfy

$$H_{\text{atom}}|i\rangle = E_i|i\rangle, \tag{7.4}$$

where E_i is the energy eigenvalue of the ith state, and the atom transition operator is

$$\sigma_{ij} = |j\rangle\langle i|. \tag{7.5}$$

The *dipole interaction* Hamiltonian between the electron at position $\mathbf{r}$ with charge $-e$ and the quantized uniform electric field $\mathbf{E}$ is

$$H_{\text{int}} = -e(\mathbf{r} \cdot \mathbf{E}) \tag{7.6}$$

and the electric–dipole transition matrix between state $|i\rangle$ and $|j\rangle$ is

$$\mathbf{p}_{ji} = e\langle j|\mathbf{r}|i\rangle. \tag{7.7}$$

Following an established notation [1], the Hamiltonian for the electron states in the atom may be written

$$H_{\text{atom}} = \sum_i E_i |i\rangle\langle i| = \sum_i E_i \sigma_{ii} \tag{7.8}$$

and, making use of equations (7.7) and (7.5),

$$e\mathbf{r} = \sum_{i,j} e|i\rangle\langle i|\mathbf{r}|j\rangle\langle j| = \sum_{i,j} \mathbf{p}_{ji}\sigma_{ji}. \tag{7.9}$$

Placing the atom in the position $\mathbf{r} = 0$ in equation (7.2) gives

$$\mathbf{E}(\mathbf{r}) = \sum_{\mathbf{k}} \mathbf{e}_{\sim\mathbf{k}} \sqrt{\frac{\hbar\omega_{\mathbf{k}}}{2\varepsilon_0 V_{\text{vol}}}} \left(\hat{b}_{\mathbf{k}} + \hat{b}_{\mathbf{k}}^{\dagger}\right) \tag{7.10}$$

so that the Hamiltonian may be written

$$H = H_{\text{atom}} + H_{\text{field}} - e(\mathbf{r} \cdot \mathbf{E})$$

$$= \sum_i E_i \sigma_{ii} + \sum_{\mathbf{k}} \hbar\omega_{\mathbf{k}}\left(\hat{b}_{\mathbf{k}}^{\dagger}\hat{b}_{\mathbf{k}} + \frac{1}{2}\right) - \sum_{i,j} \mathbf{p}_{ji}\sigma_{ji} \cdot \sum_{\mathbf{k}} \mathbf{e}_{\sim\mathbf{k}} \sqrt{\frac{\hbar\omega_{\mathbf{k}}}{2\varepsilon_0 V_{\text{vol}}}} \left(\hat{b}_{\mathbf{k}} + \hat{b}_{\mathbf{k}}^{\dagger}\right). \tag{7.11}$$

Introducing

$$g_{\mathbf{k}}^{ji} = -\frac{\mathbf{p}_{ji} \cdot \mathbf{e}_{\sim\mathbf{k}} \sqrt{\dfrac{\hbar\omega_{\mathbf{k}}}{2\varepsilon_0 V_{\text{vol}}}}}{\hbar} \tag{7.12}$$

and removing the zero-point energy[1]

$$H = \sum_i E_i \sigma_{ii} + \sum_{\mathbf{k}} \hbar\omega_{\mathbf{k}}\hat{b}_{\mathbf{k}}^{\dagger}\hat{b}_{\mathbf{k}} + \sum_{i,j}\sum_{\mathbf{k}} \hbar g_{\mathbf{k}}^{ji}\sigma_{ji}\left(\hat{b}_{\mathbf{k}} + \hat{b}_{\mathbf{k}}^{\dagger}\right). \tag{7.13}$$

7.1.2 The two-level system in the rotating wave approximation

A two-level system considers an atom with just two electron eigenstates $|1\rangle$ and $|2\rangle$ and corresponding eigenenergies E_1 and E_2. The electric–dipole transition matrix element between the two states is $\mathbf{p}_{21} = e\langle 2 | \mathbf{r} | 1\rangle = e\langle 1 | \mathbf{r} | 2\rangle = \mathbf{p}_{12}$ and so $g_{\mathbf{k}}^{21} = g_{\mathbf{k}}^{12} = g_{\mathbf{k}}$. The Hamiltonian in equation (7.13) becomes

$$H = E_1\sigma_{11} + E_2\sigma_{22} + \sum_{\mathbf{k}} \hbar\omega_{\mathbf{k}}\hat{b}_{\mathbf{k}}^{\dagger}\hat{b}_{\mathbf{k}} + \sum_{\mathbf{k}} \hbar g_{\mathbf{k}}(\sigma_{21} + \sigma_{12})\left(\hat{b}_{\mathbf{k}} + \hat{b}_{\mathbf{k}}^{\dagger}\right). \tag{7.14}$$

---

[1] Note that the sum over zero-point energy is infinite!

Completeness of the atomic energy eigenstates requires

$$\sum_i |i\rangle\langle i| = \sum_i \sigma_{ii} = 1 \tag{7.15}$$

so that for the two-level atom $\sigma_{11} + \sigma_{22} = 1$. The separation in energy eigenvalues of the atom is $\hbar\omega = E_2 - E_1$ and hence the first two terms in the Hamiltonian in equation (7.14) may be written as

$$E_1\sigma_{11} + E_2\sigma_{22} = \frac{1}{2}((E_2 - E_1)(\sigma_{22} - \sigma_{11}) + (E_2 + E_1)(\sigma_{22} + \sigma_{11}))$$
$$= \frac{\hbar\omega}{2}(\sigma_{22} - \sigma_{11}) + \frac{1}{2}(E_2 + E_1). \tag{7.16}$$

The constant energy $(E_2 + E_1)/2$ may be removed, and the Hamiltonian becomes

$$H = \frac{\hbar\omega}{2}(\sigma_{22} - \sigma_{11}) + \sum_k \hbar\omega_k \hat{b}_k^\dagger \hat{b}_k + \sum_k \hbar g_k(\sigma_{21} + \sigma_{12})\left(\hat{b}_k + \hat{b}_k^\dagger\right). \tag{7.17}$$

Creation and annihilation of the atom's two-level electron states follow the Pauli spin-half matrices

$$\sigma_+ = \begin{bmatrix} 0 & 1 \\ 0 & 0 \end{bmatrix} = \sigma_{21} = |2\rangle\langle 1| \tag{7.18}$$

$$\sigma_- = \begin{bmatrix} 0 & 0 \\ 1 & 0 \end{bmatrix} = \sigma_{12} = |1\rangle\langle 2| \tag{7.19}$$

$$\sigma_z = \begin{bmatrix} 1 & 0 \\ 0 & -1 \end{bmatrix} = \sigma_{22} - \sigma_{11} = |2\rangle\langle 2| - |1\rangle\langle 1| \tag{7.20}$$

with commutator relations

$$[\sigma_-, \sigma_+] = -\sigma_z \tag{7.21}$$

and

$$[\sigma_-, \sigma_z] = 2\sigma_-. \tag{7.22}$$

Using this notation, the Hamiltonian is

$$H = \frac{\hbar\omega}{2}\sigma_z + \sum_k \hbar\omega_k \hat{b}_k^\dagger \hat{b}_k + \sum_k \hbar g_k(\sigma_+ + \sigma_-)\left(\hat{b}_k + \hat{b}_k^\dagger\right). \tag{7.23}$$

In the rotating wave approximation, only the energy-conserving processes are retained in $(\sigma_+ + \sigma_-)(\hat{b}_k + \hat{b}_k^\dagger)$ to give $(\sigma_+\hat{b}_k + \hat{b}_k^\dagger\sigma_-)$ so that

$$H = \frac{\hbar\omega}{2}\sigma_z + \sum_k \hbar\omega_k \hat{b}_k^\dagger \hat{b}_k + \sum_k \hbar g_k\left(\sigma_+\hat{b}_k + \hat{b}_k^\dagger\sigma_-\right). \tag{7.24}$$

Removing the sum over different plane waves used to describe the photon field and considering just one plane wave that has radial frequency ω in resonance with the electron transition energy, equation (7.24) can be written to give a single-atom system Hamiltonian:

$$H_S = H_0 + H_1 = \frac{\hbar\omega}{2}\sigma_z + \hbar\omega\hat{b}^\dagger\hat{b} + \hbar g(\sigma_+\hat{b} + \hat{b}^\dagger\sigma_-), \qquad (7.25)$$

where the noninteracting Hamiltonian that consists of the atom and field is

$$H_0 = \frac{\hbar\omega}{2}\sigma_z + \hbar\omega\hat{b}^\dagger\hat{b} \qquad (7.26)$$

and the interacting term coupling the atom to the field is

$$H_1 = \hbar g(\sigma_+\hat{b} + \hat{b}^\dagger\sigma_-). \qquad (7.27)$$

7.2 The mesoscale laser

Just as micro-LEDs (contemporary LEDs scaled to dimensions of a few microns) have created new applications in communications and visible display technology [2], it seems appropriate to study limits to reducing the size of semiconductor laser diodes to potentially increase the number and type of applications for which they may be used [3]. Efficient and very small laser diode photon sources could contribute to removing communication bottlenecks in processor architectures [4] and, as with micro-LEDs, find applications in displays. To explore these and other potential uses of small (mesoscale) lasers, developing a model that is accurate enough to guide analysis can be helpful.

In chapter 6, semiclassical master equations using continuum probability functions describe the occupation of quantized particle number states. This approach successfully accounts for energy and particle number conservation in a mesoscale laser. The model illustrates the critical role of quantum fluctuations in determining a mesoscale laser's steady-state and transient response [5]. Key predictions of this minimalist model are that fluctuations can suppress lasing near the threshold and that bimodal, non-Poisson, probability distributions in photon number and excited electronic state number can occur. However, phase information and associated correlations are missing from a model that only considers quantized particle number states. Addressing the physical presence of phase and associated correlation effects requires including a more complete approach to quantized atom and photon states.

An externally excited atom or multiple atoms coupled to a photon mode of a cavity can be configured as a laser. Semiconductor quantum dots may be considered as artificial atoms in a laser diode. Such atoms are driven from equilibrium and excited by current from a source such as a battery. In general, the electronic states of the atom emitters are pumped at a rate P_r. De-excitation of emitter atom electron states can be via photon emission.

At this level of abstraction, the mesoscale laser may be considered as consisting of a non-zero positive integer N_e two-level atom or quantum dot emitters that are incoherently pumped by an external reservoir. The atoms interact with a quantized photon field of positive integer S_n photons in the optical cavity. The photons in the cavity can decay into an external reservoir by transmission through the finite reflectivity mirrors of the optical cavity.

The Hamiltonian describing N_e two-level emitters in a single quantum dot laser [6] or mesoscale laser [7] may be expressed as simply

$$H = H_S + H_R + H_{RS}, \tag{7.28}$$

where the Hamiltonian of the system of emitters and cavity photons is H_S, the Hamiltonian for the reservoirs is H_R, and the system is coupled to the reservoirs by H_{RS}. For a homogeneous system of identical emitters, the Jaynes–Cummings Hamiltonian [8] in the rotating wave approximation (equation (7.25)) coupling a single-cavity optical mode with a sum over N_e two-level emitters can be used for H_S. An external reservoir incoherently pumps the emitter electronic states at a constant rate P_r. The ground $|1\rangle$ and excited $|2\rangle$ electronic states of each emitter have eigenenergy E_1 and E_2, respectively. The stimulated and spontaneous emission coefficient of each emitter is g. The angular frequency of the necessarily high-Q single optical cavity resonance may be set to ω such that $\hbar\omega = E_2 - E_1$. The laser photon field at this frequency can decay into the single-cavity mode by coupling to an external reservoir via partially transmitting mirrors characterized by a total photon loss rate of κ. Spontaneous emission into nonlasing leaky optical modes acts to dampen the excited states of the emitters at a rate of γ. To avoid the contribution of thermal fluctuations, the absolute temperature of the system is set to zero.

The Hamiltonian given by equation (7.28) can be solved for N_e two-level emitters coupled to a single optical mode as a function of incoherent pump rate, P_r [6, 7]. To compare results for different numbers of emitters, it is helpful to define a normalized pump rate such that $P_{norm} = P_r/N_e$. Figure 7.1 shows typical results of calculating the average photon number $\langle S_n \rangle$ in the lasing mode, the average net inversion of emitter states, the relative variance (Fano-factor) of photon fluctuations, $(\langle S_n^2 \rangle - \langle S_n \rangle^2)/\langle S_n \rangle = \sigma_{S_n}^2/\langle S_n \rangle$, and the spectral linewidth $\Delta\omega_{FWHM}$ of photon emission into the laser mode as a function of normalized pump rate using a $\log_{10}$ scale [9]. Notice that all rate parameters are in units of meV.

As might be anticipated for such a small system containing just a few emitters, with increasing pump rate there is no well-defined change in average photon emission $\langle S_n \rangle$ that can be assigned to the laser threshold. However, the transition between characteristic operating regimes of the system has a peak in photon fluctuations and, hence, a peak in photon Fano-factor. The low value of pump rate at which this peak occurs can be used to define the laser threshold. In figure 7.1(c), a broad peak in Fano-factor associated with lasing threshold occurs when $P_{norm} < 1$ meV. At higher values of pump rate, in this case $P_{norm} > 10$ meV, figure 7.1 shows a peak in $\langle S_n \rangle$, saturation in emitter inversion [10], a peak in the Fano-factor, and an increase in $\Delta\omega_{FWHM}$. These are all self-quenching signatures, an effect not predicted by the semiclassical master equations developed in chapter 6. The peaks in photon Fano-factor shown in figure 7.1(c) are fundamentally related to the fluctuation–dissipation theorem [11].

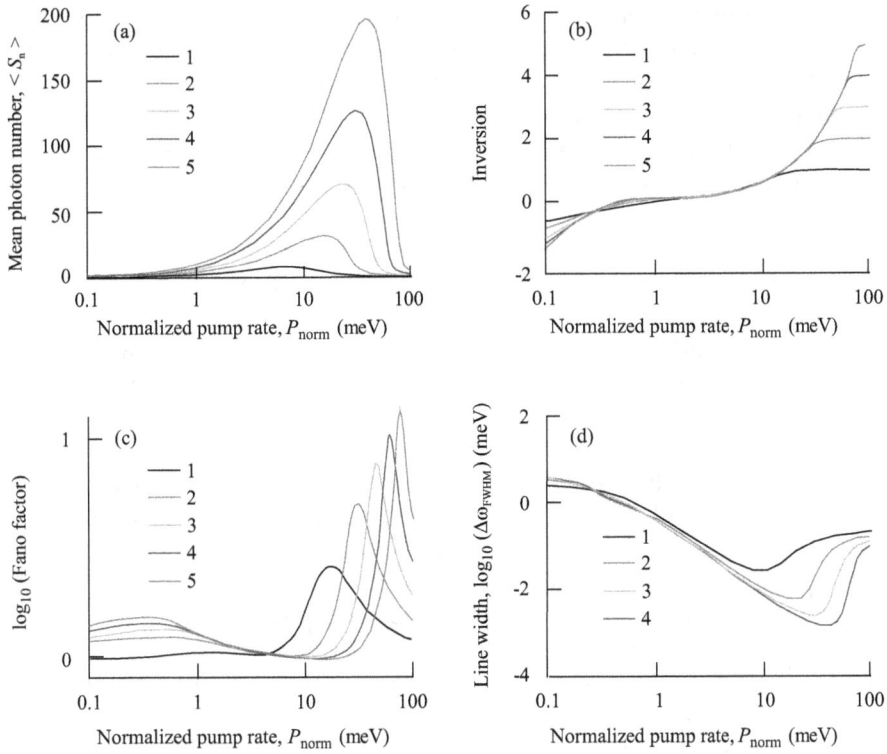

Figure 7.1. Steady-state properties for different numbers of two-level emitters coupled to a cavity field with normalized incoherent pump rate P_{norm} on a $\log_{10}$ scale and calculated using the Hamiltonian given by equation (7.28). (a) Average photon number in lasing mode, $\langle S_n \rangle$. (b) Average net inversion of emitters. (c) Photon Fano-factor. (d) Time-averaged spectral linewidth, $\Delta\omega_{FWHM}$, in the photon mode. Calculations use parameters $g = 1$ meV, $\gamma = 0.1$ meV, $\omega = 1000$ meV, and $\kappa = 0.25$ meV. (Adapted with permission from [9]. Copyright 2019 IEEE.)

The FWHM spectral linewidth $\Delta\omega_{FWHM}$ of photon emission from a macroscopic semiconductor laser diode is predicted by equation (5.28) to be inversely proportional to the average photon number in the lasing mode. If, as is usually the case for a conventional macroscopic device above the lasing threshold, $\langle S_n \rangle \propto P_r$, then $\Delta\omega_{FWHM} \propto 1/P_r$. The behavior in a mesolaser can be different. The average photon number in the lasing mode plotted in figure 7.1(a) is not linearly proportional to the pump rate, and so spectral linewidth is not inversely proportional to the pump rate. However, $\Delta\omega_{FWHM}$ does initially decrease with increasing pump rate in the mesoscale laser. At larger pump rates, the system approaches self-quenching, and quantum state fluctuations increase photon emission spectral linewidth.

The peak value of the average photon number $\langle S_n \rangle$ due to self-quenching increases as N_e^2 and the corresponding peak value in the Fano-factor increases as N_e. The Fano-factor peak associated with the lasing threshold ($P_{norm} < 1$ meV) increases as $\sqrt{N_e}$.

As with the experiments described in section 5.3, a mesoscale laser can be compared to a nonlasing device by determining system characteristics as cavity mirror losses are increased. In a device with relatively small $\kappa = 0.25$ meV, sub-threshold photon fluctuations into the lasing mode and inversion pinning above threshold are signatures of lasing behavior. These characteristics are entirely absent for a device in which the optical cavity has been effectively removed (for example, $\kappa = 25$ meV).

Figure 7.2 shows the results of calculating average net inversion for $N_e = 3$ emitters in an optical cavity as a function of normalized pump rate, P_{norm}, for the indicated values of photon mode loss rate κ. The key result illustrated in figure 7.2 is the emergence of emitter average inversion pinning with decreasing cavity loss rate, κ. This occurs because of increased photon number and stimulated emission as the photon loss rate is reduced.

Figure 7.3 shows the calculated steady-state average net inversion as a function of pump rate P_r for the indicated number of two-level emitters coupled to a cavity field. Note the *linear* scale of P_r. The threshold pump rate (as measured by the peak in curvature of inversion as a function of pump rate) increases approximately linearly with N_e because interaction with the incoherent pump is assumed to be shared equally between emitters. Inversion pinning becomes more effective as N_e increases, and the above threshold slope of average net inversion scales as $1/N_e$.

Mesoscale lasers, consisting of a few emitters in a dissipative quantum system, have internal particle dynamics subject to strong quantum fluctuations whose effect is to partially suppress the classical lasing behavior found in a conventional macroscopic semiconductor laser diode. Relatively weak optical gain in a mesoscale

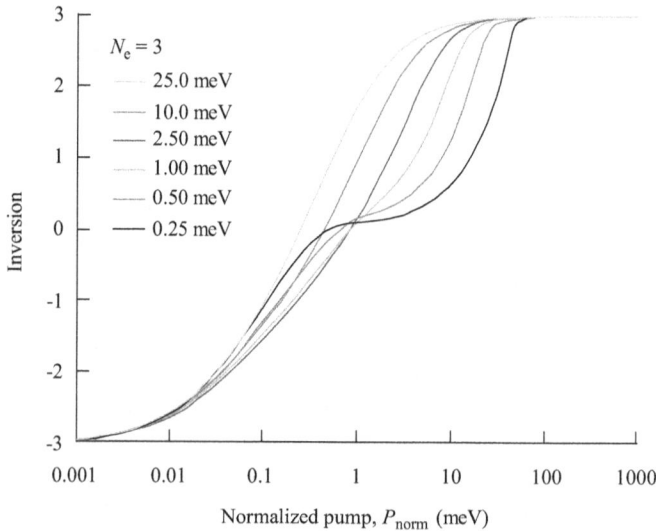

Figure 7.2. Calculated average net inversion of emitters as a function of normalized pump rate for the indicated values of κ. Parameters are $N_e = 3$, $g = 1$ meV, $\gamma = 0.1$ meV, $\omega = 1000$ meV. As the photon cavity loss rate decreases, the inversion pinning characteristic of a laser emerges.

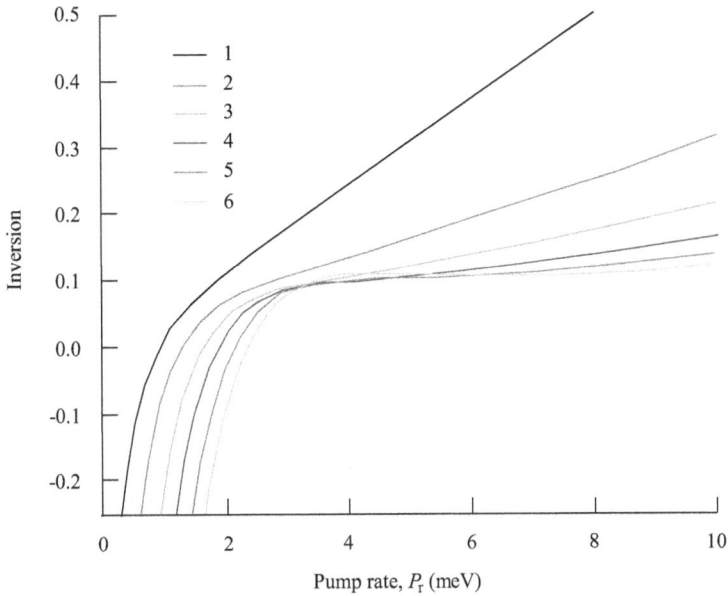

Figure 7.3. Calculated steady-state average net inversion of the indicated number of two-level emitters coupled to a cavity field on resonance as a function of incoherent pump rate P_r. The parameters are $g = 1$ meV, $\gamma = 0.1$ meV, $\omega = 1000$ meV, $\kappa = 0.25$ meV.

laser, photon cavity losses, and the presence of quantum fluctuations remove the abrupt change in slope of $\langle S_n \rangle$ with increasing pump rate around the laser threshold. However, a broad peak in photon fluctuations near the threshold is retained.

Studying the classical-to-quantum transition in a scaled semiconductor laser can reveal how the nonequilibrium phase transition characterizing a conventional macroscopic device is obscured by quantum fluctuations in a mesoscale device. While mesoscale lasers cannot behave in the same way as large devices containing many emitters and operating in the large particle number limit (the thermodynamic limit), they are able to retain key signatures of lasing such as inversion pinning and a peak in photon fluctuations near threshold pump rate. Changing experimentally accessible control parameters, such as decreasing photon loss rates κ and γ, and increasing N_e to enhance gain, can be used to adjust the behavior of a mesoscale laser to approach that of a macroscopic device. However, in mesoscale laser device designs that specify a low number of emitters, quantum mechanical self-quenching will always limit photon emission intensity at high pump rates.

The emergence of long-lived emitter states is predicted to occur in a mesolaser driven below the self-quenching threshold. Distinct behavioral regimes are separated by peaks in photon fluctuations (the Fano-factor) as a function of pump rate in a manner analogous to but fundamentally different from the nonequilibrium phase transition analogy in conventional lasers (see chapter 5).

There is a richness in solutions to the Hamiltonian given by equation (7.28) that is not present in the semiclassical master equations described in chapter 6. One such

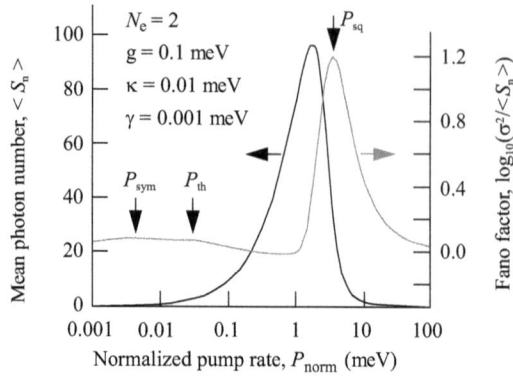

Figure 7.4. Steady-state average photon number $\langle S_n \rangle$ and photon Fano-factor for a system with $N_e = 2$ two-level emitters coupled to a cavity field with an incoherent pump rate P_{norm} plotted using a $\log_{10}$ scale. The indicated peaks in Fano-factor at the normalized pump rate correspond to the normalized laser threshold (P_{th}), self-quenching (P_{sq}), and destruction of the symmetry-protected state (P_{sym}), respectively. (Adapted with permission from [9]. Copyright 2019 IEEE.)

example is the existence of long-lived emitter states at low pump rates. These are special symmetry-protected entangled quantum states [12] consisting of more than one emitter whose emission into the cavity mode is suppressed. Such long-lived (almost) dark states have a lifetime that is reduced with increasing pump rate. In this way, the system can transition between different operational characteristics in which the lifetime of states changes [13]. As it transitions, there is dissipation, fluctuations, and a peak in photon Fano-factor at normalized pump value P_{sym} as a class of dark states are no longer supported. Peaks in photon Fano-factor appear as a function of normalized pump rate, and these are shown in figure 7.4 for the case when $N_e = 2$ [9]. The peaks occur at normalized pump rates P_{sym}, P_{th}, and P_{sq} for symmetric dark states, the laser threshold, and self-quenching, respectively. In general, the boundary between different dominant modes of operation is associated with fluctuations and dissipation. Long-lived dark states prevail when $P_{norm} < P_{sym}$. When $P_{sym} < P_{norm} < P_{th}$ spontaneous emission and sub-threshold fluctuations occur, and when $P_{th} < P_{norm} < P_{sq}$ lasing behavior dominates. Increasing the pump rate such that $P_{norm} > P_{sq}$ causes a peak in photon fluctuations as the system transitions to the self-quenching regime.

7.3 Beyond the mesoscale laser

A typical laser diode is a macroscopic device for which the analogy with a classical nonequilibrium second-order phase transition is valid. If the laser diode is made smaller (scaled), it can enter the mesoscale regime in which the classical description of device behavior becomes less valid, and a more quantum description is appropriate. In this classical-to-quantum transition, many of a conventional laser diode's known commercially valuable properties are lost. The new quantum properties that emerge have yet to be effectively controlled, and compelling applications have yet to be found.

Recognizing the practical limitations of mesoscale lasers, there is interest in other device configurations and exploring their associated potential applications. For example, conditions in which photons and matter interact strongly can result in new and exciting phenomena. It is possible to use strong electromagnetic fields to create coherent populations of atom states, which can lead to remarkable nonlinear effects. A well-known example is *self-induced transparency*, in which a short coherent optical pulse above a critical input energy can pass through an optically resonant medium of two-level atoms as if it were transparent. Below a critical energy, the optical pulse is absorbed in the medium [14].

Systems of *three-level* atoms can be prepared so that a superposition of ground-state doublets cancels absorption, ensuring that there are more atoms in the ground-state doublet than in the excited state. These facts may be used to claim *lasing without inversion* [15]. Refractive index enhancement is also achievable in a system of three-level atoms when low-lying doublet atomic states are coherent. In this case, a large refractive index with zero loss is, in principle, achievable [16].

All of these and other aspects of light–matter interaction remain interesting. However, finding compelling, commercially viable, practical applications that exploit these quantum phenomena is an exciting but extraordinarily difficult and outstanding challenge.

Bibliography

[1] Scully M O and Zubairy M S 1997 *Quantum Optics* (Cambridge: Cambridge University Press)
[2] For example, Singh K J *et al* 2023 *J. Lightwave Technol.* **41** 1480
[3] For reviews, see Deng H, Lippi G L, Mørk J, Wiersig J and Reitzenstein S 2021 *Adv. Opt. Mat.* **19** 2100415
Ren K, Li C, Fang Z and Feng F 2023 *Laser Phot. Rev.* **17** 2200758
[4] Levi A F J 2000 *Proc. IEEE* **88** 750
[5] Roy-Choudhury K and Levi A F J 2009 *Phys. Rev. Lett.* **102** 053902
[6] Perea J I, Porras D and Tejedor C 2004 *Phys. Rev.* B **70** 115304
[7] Roy-Choudhury K and Levi A F J 2011 *Phys. Rev.* A **83** 043827
[8] Jaynes E T and Cummings F W 1963 *Proc. IEEE* **51** 89
[9] Levi A F J 2019 Photonics and Electromagnetics Research Symposium – Spring (PIERS-Spring) *(Rome, Italy)* p 684
[10] Atlasov K *et al* 2010 Saturation in laser emission is observed with increasing pump power in quantum-wire photonic-crystal microcavity lasers *Opt. Lett.* **35** 1154
[11] Kubo R 1966 *Rep. Prog. Phys.* **29** 255
[12] Zanardi P 1997 *Phys. Rev.* A **56** 4445
[13] Abouzaid A, Unglaub W and Levi A F J 2019 *J. Phys.* B: At. Mol. Opt. Phys. **52** 245401
[14] McCall S L and Hahn E L 1967 *Phys. Rev. Lett.* **18** 908
[15] Nottelmann A, Peters C and Lange W 1993 *Phys. Rev. Lett.* **70** 1783
[16] Scully M O 1991 *Phys. Rev. Lett.* **67** 1855

Appendix A

Physical values

SI-MKS. See http://physics.nist.gov/constants.

Speed of light in free space $c = 1/\sqrt{\varepsilon_0 \mu_0}$	$c = 2.997\ 924\ 58 \times 10^8\ \text{m s}^{-1}$
Planck's constant	$\hbar = 1.054\ 571\ 817\ \dots \times 10^{-34}\ \text{J s}$
Electron charge	$e = 1.602\ 176\ 634 \times 10^{-19}\ \text{C}$
Electron mass	$m_0 = 9.109\ 383\ 7015(28) \times 10^{-31}\ \text{kg}$
Neutron mass	$m_n = 1.674\ 927\ 498\ 05(70) \times 10^{-27}\ \text{kg}$
Proton mass	$m_p = 1.672\ 621\ 923\ 69(51) \times 10^{-27}\ \text{kg}$
Boltzmann constant	$k_B = 1.380\ 649 \times 10^{-23}\ \text{J K}^{-1}$
Permittivity of free space	$\varepsilon_0 = 8.854\ 187\ 8128(13) \times 10^{-12}\ \text{F m}^{-1}$
Permeability of free space	$\mu_0 = 4\pi \times 10^{-7}\ \text{H m}^{-1}$
Avogadro's number	$N_A = 6.022\ 140\ 76 \times 10^{23}\ \text{mol}^{-1}$
Bohr radius	$a_B = 5.291\ 772\ 109\ 03(80) \times 10^{-11}\ \text{m}$
$a_B = \dfrac{4\pi\varepsilon_0 \hbar^2}{m_0 e^2}$, effective Bohr radius	$a_B^* = \dfrac{4\pi\varepsilon_0 \varepsilon_{r0} \hbar^2}{m_e^* e^2}$
Inverse fine-structure constant $\alpha_f^{-1} = \dfrac{4\pi\varepsilon_0 \hbar c}{e^2}$	$\alpha_f^{-1} = 137.035\ 999\ 084(21)$

doi:10.1088/978-0-7503-6417-1ch8   A-1

IOP Publishing

Essential Semiconductor Laser Device Physics (Second Edition)

A F J Levi

Appendix B

Crystal structure

B.1 Crystal structure

Three-dimensional crystals may be described as a periodic array of atoms located in space on a *lattice*. A fixed group of atoms called a *basis* is associated with each lattice point. The crystal is exactly filled when basis atoms are placed at each lattice point. The orientation of basis atoms does not change at each lattice point. For a given lattice, there exists a primary *unit cell* that can be defined by the three vectors $\mathbf{a}_1$, $\mathbf{a}_2$, and $\mathbf{a}_3$. Crystal structure is the real-space translation of basic points throughout space via

$$\mathbf{R} = n_1\mathbf{a}_1 + n_2\mathbf{a}_2 + n_3\mathbf{a}_3, \tag{B.1}$$

where n_1, n_2, and n_3 are integers [1]. This complete real-space lattice is called the *Bravais lattice*. The volume of the smallest unit cell that can be used to form the lattice is called a *primitive cell* and is

$$\Omega_{\text{cell}} = \mathbf{a}_1 \cdot (\mathbf{a}_2 \times \mathbf{a}_3). \tag{B.2}$$

There is no unique way of choosing a primitive cell. Often, a *Wigner–Seitz* primitive cell is used. In this case, a lattice reference point is specified such that any point of the cell is closer to that lattice point than any other. The Wigner–Seitz primitive cell may be found by determining the smallest volume enclosed after bisecting with perpendicular planes all vectors connecting a reference atom to all-atom positions in the crystal. A Bravais lattice can also be described using a non-primitive unit cell. For example, there is a conventional unit cell used for crystals with cubic symmetry that is not a primitive unit cell.

As illustrated in figure B.1, the *cell parameters* of a three-dimensional crystal can be defined in terms of vector lengths $|\mathbf{a}_1|$, $|\mathbf{a}_2|$, and $|\mathbf{a}_3|$ and angles α, β, γ. In the figure the corresponding unit-normal vectors are $\underset{\sim}{\mathbf{a}}_1$, $\underset{\sim}{\mathbf{a}}_2$, and $\underset{\sim}{\mathbf{a}}_3$, respectively. The interaxial angles are defined as $\mathbf{a}_1 \wedge \mathbf{a}_2 = \gamma$, $\mathbf{a}_1 \wedge \mathbf{a}_3 = \beta$, and $\mathbf{a}_2 \wedge \mathbf{a}_3 = \alpha$.

doi:10.1088/978-0-7503-6417-1ch9
B-1

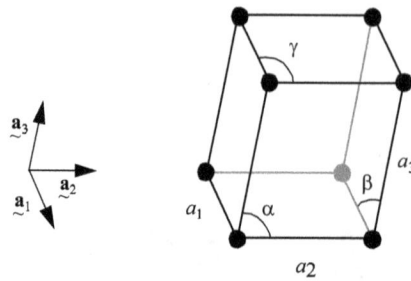

Figure B.1. Illustration of cell parameters for a crystal in three dimensions consisting of lengths a_1, a_2, a_3, and angles α, β, and γ.

Table B.1. Crystal systems in three dimensions.

Crystal system	Number of lattices	Cell parameter constraints
Triclinic	1	$a_1 \neq a_2 \neq a_3$, $\alpha \neq \beta \neq \gamma$
Monoclinic	2	$a_1 \neq a_2 \neq a_3$, $\alpha = \gamma = 90° \neq \beta$
Orthorhombic	4	$a_1 \neq a_2 \neq a_3$, $\alpha = \beta = \gamma - 90°$
Tetragonal	2	$a_1 = a_2 \neq a_3$, $\alpha = \beta = \gamma - 90°$
Cubic	3	$a_1 = a_2 = a_3$, $\alpha = \beta = \gamma - 90°$
Trigonal	1	$a_1 = a_2 = a_3$, $\alpha = \beta = \gamma < 120°$, $\neq 90°$
Hexagonal	1	$a_1 = a_2 \neq a_3$, $\alpha = \beta = 90°$, $\gamma - 120°$

There are seven crystal systems in three dimensions listed in table B.1. The lowest symmetry system is triclinic for which there are no restrictions on the values of the cell parameters. In the other crystal systems, symmetry reduces the number of unique lattice parameters.

Figure B.2 shows the 14 different point lattices that exist in three dimensions. There are three-point lattices for the cubic crystal system. They are simple cubic (SC), body-centered cubic (BCC), and face-centered cubic (FCC).

B.2 Cubic lattices in three dimensions

Figure B.3 shows the conventional unit cell for three types of cubic crystal structure, SC, BCC, and FCC, with *lattice constant L*.

The FCC unit cell is the same as the SC but with an additional atom on each face of the cube. Elements with FCC crystal structure include Al ($L = 0.405$ nm), Ni ($L = 0.352$ nm), Au ($L = 0.408$ nm), Cu ($L = 0.361$ nm), Ag, Pd, Pb, and Pt.

B.3 Diamond and zinc blende crystal structure

The diamond crystal structure is cubic and can be constructed using a lattice consisting of *two* identical interpenetrating FCC unit cells offset from each other by

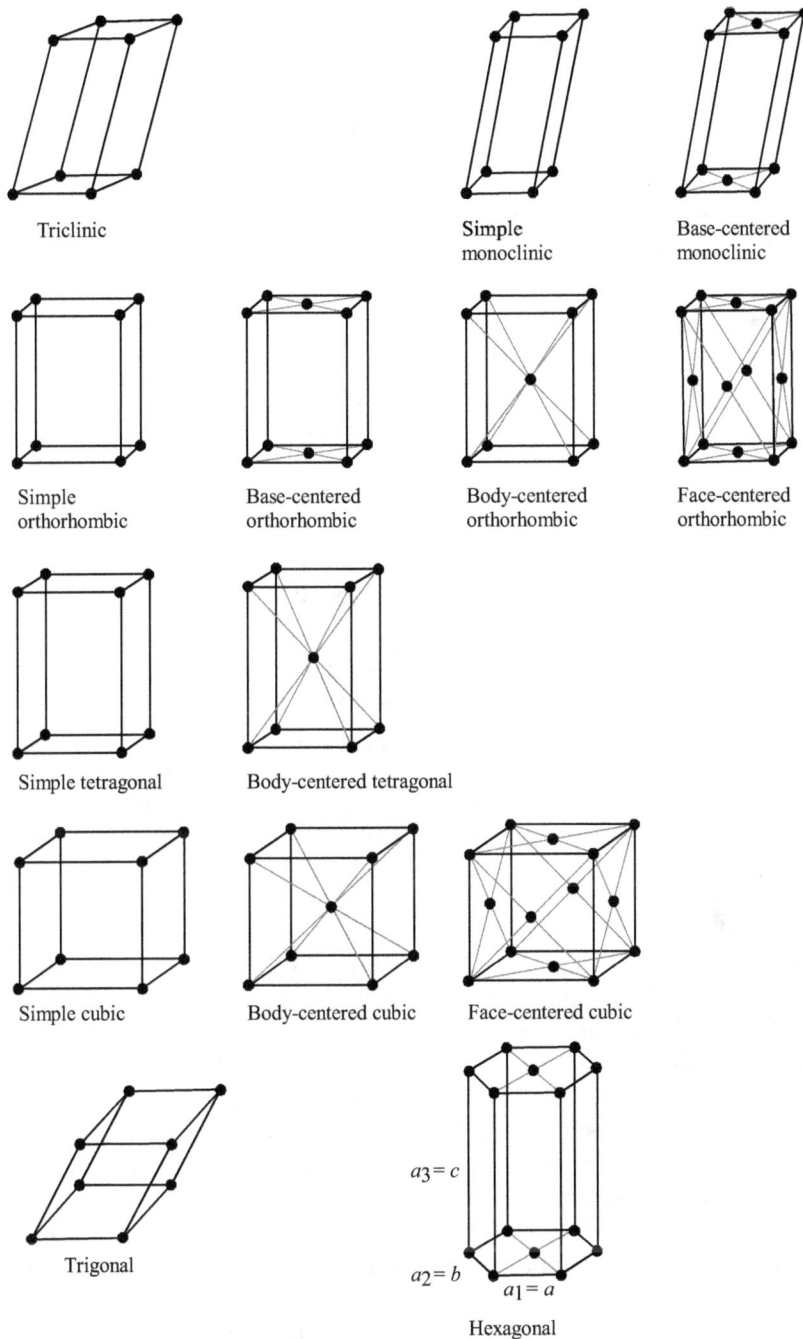

Triclinic

Simple
monoclinic

Base-centered
monoclinic

Simple
orthorhombic

Base-centered
orthorhombic

Body-centered
orthorhombic

Face-centered
orthorhombic

Simple tetragonal

Body-centered tetragonal

Simple cubic

Body-centered cubic

Face-centered cubic

Trigonal

$a_3 = c$

$a_2 = b$

$a_1 = a$

Hexagonal

Figure B.2. Sketches of the 14 different point lattices in three dimensions.

Simple cubic (SC)   Body-centered cubic (BCC)   Face-centered cubic (FCC)

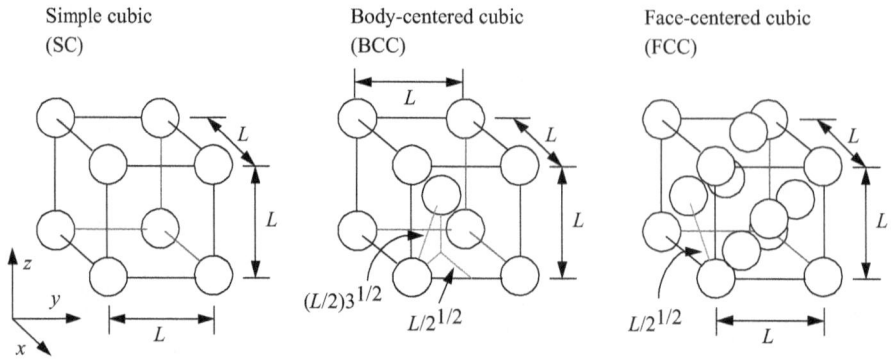

Figure B.3. Illustration of the indicated three-dimensional cubic conventional unit cells, each of lattice constant L. In the figure, each sphere represents the position of an atom.

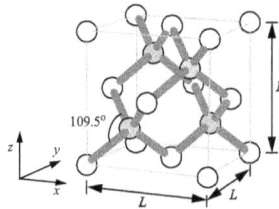

Figure B.4. Conventional cubic cell of the zinc blende lattice consists of two interpenetrating FCC Bravais lattices offset by a quarter body diagonal. For the semiconductor GaAs, atom type alternates between Ga (dark spheres) and As (white spheres) and the crystal lattice constant is $L = 0.565$ nm. Nearest neighbor bonds are shown as thick lines. Each atom is tetrahedrally coordinated with its nearest neighbor. There is a 109.5° angle between adjacent bonds.

a quarter body diagonal, $L(\mathbf{x} + \mathbf{y} + \mathbf{x})/4$. The diamond crystal lattice has a *basis* of two *identical* atoms. The diamond lattice is not a Bravais lattice because the orientation of tetrahedrally coordinated neighbors is rotated. Since there are four atoms per FCC conventional unit cell, the diamond unit cell contains $2 \times 4 = 8$ atoms. Elements with a *diamond crystal structure* include Si ($L = 0.543$ nm), Ge ($L = 0.566$ nm), and C ($L = 0.357$ nm) [2]. It follows, for example, that a silicon crystal has about 5×10^{22} silicon atoms per cubic centimeter.

The zinc blende crystal lattice is identical to diamond but with a basis of two *different* atoms. Materials with the zinc blende crystal structure include the group III–V compound semiconductors GaAs, AlAs, InP, InAs, InSb, InGaAs, and InGaAsP. Many of these compound semiconductors have direct band gaps and find application in optoelectronic devices such as laser diodes and photodetectors. Figure B.4 illustrates the conventional zinc blende unit cell for GaAs.

B.4 Hexagonal crystal structure

Some group II–VI and III–V compound semiconductors have the Wurtzite crystal structure. This is an example of the hexagonal crystal system. Compound semiconductors with Wurtzite crystal structure include ZnO, ZnSe, CdS, CdSe, and GaN. These semiconductors can also be crystallized in the zinc blende structure.

B.5 The reciprocal lattice

Given the basic unit cell defined by the vectors $\mathbf{a}_1$, $\mathbf{a}_2$, and $\mathbf{a}_3$ in real space, three fundamental reciprocal vectors, $\mathbf{g}_1$, $\mathbf{g}_2$, and $\mathbf{g}_3$, may be constructed in reciprocal space such that $\mathbf{a}_i \cdot \mathbf{g}_j = 2\pi\delta_{ij}$, with $\mathbf{g}_1 = 2\pi(\mathbf{a}_2 \times \mathbf{a}_3)/\Omega_{\text{cell}}$, $\mathbf{g}_2 = 2\pi(\mathbf{a}_3 \times \mathbf{a}_1)/\Omega_{\text{cell}}$, and $\mathbf{g}_3 = 2\pi(\mathbf{a}_1 \times \mathbf{a}_2)/\Omega_{\text{cell}}$.

Crystal structure may be defined as a reciprocal-space translation of basic points throughout the space, in which

$$\mathbf{G} = n_1\mathbf{g}_1 + n_2\mathbf{g}_2 + n_3\mathbf{g}_3, \tag{B.3}$$

where n_1, n_2, and n_3 are integers. The complete space spanned by $\mathbf{G}$ is called the *reciprocal lattice*. The volume of the three-dimensional reciprocal-space unit cell is

$$\Omega_k = \mathbf{g}_1 \cdot (\mathbf{g}_2 \times \mathbf{g}_3) = \frac{(2\pi)^3}{\Omega_{\text{cell}}}. \tag{B.4}$$

The *Brillouin zone* of the reciprocal lattice has the same definition as the Wigner–Seitz cell in real space. The first Brillouin zone is the smallest enclosed volume found by bisecting with perpendicular planes all reciprocal-lattice vectors.

Figure B.5 illustrates the first Brillouin zone with high-symmetry points labeled for a face-centered cubic lattice with lattice constant L. As may be seen, in this case, the Brillouin zone is a truncated octahedron.

The semiconductor GaAs has the zinc blende crystal structure with a basis of two atoms and, as shown in figure B.6(a), is the same as the tetrahedrally coordinated

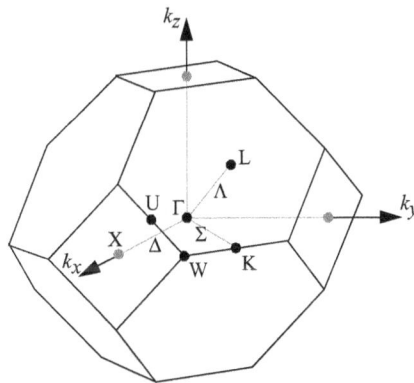

Figure B.5. Illustration of the Brillouin zone for the face-centered cubic lattice (FCC) with lattice constant L. Some high-symmetry points are $\Gamma = (0, 0, 0)$, $X = 2\pi(1, 0, 0)/L$, $L = 2\pi(0.5, 0.5, 0.5)/L$, $W = 2\pi(1, 0.5, 0)/L$. The high-symmetry line between the points Γ and X is labeled Δ, the line between the points Γ and L is Λ, and the line between Γ and K is Σ.

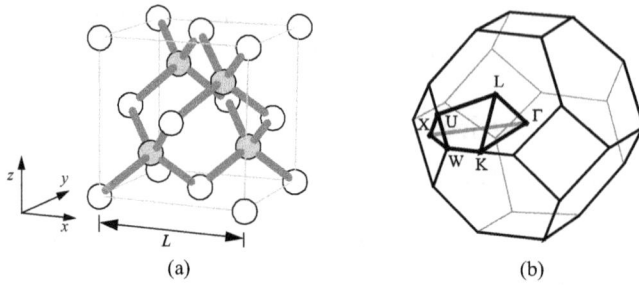

(a)   (b)

Figure B.6. (a) The real-space diamond lattice cubic conventional unit cell with lattice constant L and tetrahedrally coordinated nearest neighbor bonds are shown as thick lines. The zinc blende crystal structure of GaAs is the same as diamond, except that instead of one atom type populating the lattice, the atom type alternates between Ga (dark spheres) and As (white spheres). If the Ga atom is located at $(0, 0, 0)$ then the four nearest neighbor tetrahedrally coordinated As atoms are located at $(1, 1, 1)L/4$, $(-1, -1, 1)L/4$, $(1, -1, -1)L/4$, and $(-1, 1, -1)L/4$. (b) Reciprocal-space Brillouin zone symmetry points for the diamond lattice. There is an irreducible wedge with the vertices shown. The wedge has five faces with planes $k_x + k_y + k_z = (3/2)(2\pi/L)$, $k_x = k_y$, $k_y = k_z$, $k_z = 0$, and $k_x = 2\pi/L$, and a volume that is 1/48 of the full zone.

diamond cubic structure except that instead of one atom type populating the lattice, the atom type alternates between Ga and As. In the figure, the nearest neighbor bonds are shown as thick lines. As with the diamond structure, there are eight atoms per conventional unit cell. Group III–V compound semiconductors with the zinc blende crystal structure include GaAs ($L = 0.565$ nm), AlAs ($L = 0.566$ nm), AlGaAs, InP ($L = 0.587$ nm), InAs ($L = 0.606$ nm), InGaAs, and InGaAsP.

The Brillouin zone crystal lattice symmetry points for both diamond and zinc blende are illustrated in figure B.6(b). The *irreducible wedge*, from which the full Brillouin zone may be constructed and shown in the figure, has five faces and vertices located at Γ, L, U, X, W, K. The wedge volume is 1/48 of the full zone.

The Brillouin zone symmetry points are the following:

$$\Gamma = \quad \xi(0, 0, 0)$$
$$L = \xi(0.5, 0.5, 0.5)$$
$$X = \quad \xi(1, 0, 0)$$
$$U = \xi(1, 0.25, 0.25)$$
$$K = \xi(0.75, 0.75, 0)$$
$$W = \quad \xi(1, 0.5, 0),$$

where $\xi = 2\pi/L$. The Γ–L direction is Λ and has magnitude $\xi = \sqrt{3}/2$ at the L point. The Γ–X-direction is Δ and has magnitude ξ at the X point.

The Brillouin zone crystal lattice symmetry points in figure B.6(b) can also be shown in an extended form as illustrated in figure B.7.

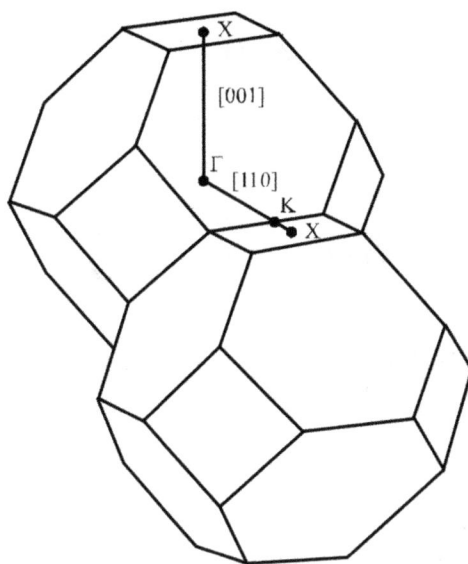

Figure B.7. Reciprocal-space Brillouin zone symmetry points for the diamond lattice illustrated in an extended scheme.

References

[1] Ashcroft N W and Mermin N D 1976 *Solid State Physics* (New York: Holt, Reinhart and Winston)
[2] Galasso F S 1970 *Structure and Properties of Inorganic Solids* (Oxford: Pergamon)

IOP Publishing

Essential Semiconductor Laser Device Physics (Second Edition)

A F J Levi

Appendix C

Tight-binding complex band structure

C.1 A single s-band in a one-dimensional lattice

For simplicity, consider a one-dimensional crystal with a primitive cell that contains one atom with lattice sites at positions $x_n = nL$, where n is an integer, and L is the nearest-neighbor atom spacing. If the potential of the nth atom at position $x_n = nL$ is $V(x - x_n)$ then the total potential for the crystal is the sum of single-atom potentials. The time-independent Schrödinger equation for an electron at position x is equation (1.18), where ψ_k is a wave function that must satisfy the Bloch condition, equation (1.23), $\psi_k(x + L) = \psi_k(x)e^{ikL}$. The Bloch condition can be satisfied using *localized basis functions* (Wannier functions) $\phi(x)$. The Wannier functions are localized around the lattice site x_n and are assumed orthogonal for different lattice points so that

$$\int \phi^*(x - x_m)\phi(x - x_n)\mathrm{d}x = \delta_{mn}. \tag{C.1}$$

The Wannier functions are related to the Bloch functions via the direct lattice sum given by equation (1.24), $\psi_{k_x}(x) = \frac{1}{\sqrt{N}}\sum_n^N e^{ik_x nL}\phi(x - nL)$.

To first order, the expectation value of electron energy is (ignoring the normalization factor $1/N$)

$$E_k = \int \psi_k^*(x)\hat{H}\psi_k(x)\mathrm{d}x = \sum_{x_m}\sum_{x_n} e^{ik(x_n - x_m)}\int \phi_k^*(x - x_m)\hat{H}\phi(x - x_n)\mathrm{d}x. \tag{C.2}$$

If there is little overlap between atomic electron wave functions that are separated by two or more nearest-neighbor atom spacings in the crystal, then only two integrals need to be kept. The first is the shifted (renormalized) atom energy levels

$$-E_0 = \int \phi_k^*(x)\hat{H}\int \phi_k(x)\mathrm{d}x \tag{C.3}$$

and the second is the contribution from overlaps of nearest neighbors

doi:10.1088/978-0-7503-6417-1ch10   C-1   © IOP Publishing Ltd 2025. All rights,

$$-t_{\text{hop},1} = \int \phi_k^*(x - x_m)\hat{H}\phi_k(x - x_n)\mathrm{d}x \qquad (\text{C.4})$$

in which $x_m = x_n \pm L$. The value of $t_{\text{hop},1}$ is determined from the overlap (or hopping) integral with the sign convention that $t_{\text{hop},1}$ is negative for s-atomic orbitals and positive for p-atomic orbitals.

The Schrödinger equation for a single s-orbital electron on the nth site with nearest-neighbor hopping may be written

$$\hat{H}\phi_n = -E_0\phi_n - t_{\text{hop},1}\phi_{n-1} - t_{\text{hop},1}\phi_{n+1} = E\phi_n. \qquad (\text{C.5})$$

Hence, because Bloch's theorem requires $\phi_{n+1} = \phi_n \mathrm{e}^{ikL}$, the electron dispersion relation for propagating states in a linear chain of such s-orbital electrons with nearest-neighbor atom spacing L is

$$E_k = -E_0 - t_{\text{hop},1}(\mathrm{e}^{ikL} + \mathrm{e}^{-ikL}) = -E_0 - 2t_{\text{hop},1}\cos(kL). \qquad (\text{C.6})$$

The cosine tight-binding band for nearest-neighbor hopping in one dimension has an energy bandwidth of $E_b = 4t_{\text{hop},1}$. Setting $E_0 = 0$ gives

$$E_k = -2t_{\text{hop},1}\cos(kL). \qquad (\text{C.7})$$

This is the $E(k)$ dispersion relation for propagating electron states in a one-dimensional lattice of atoms with nearest-neighbor interaction between s-orbitals.

If next-nearest-neighbor hopping is included, then there is an additional overlap integral $t_{\text{hop},2}$ involving sites $x_m = x_n \pm 2L$ that modifies the dispersion relation for propagating electron states to give

$$E_k = -2t_{\text{hop},1}\cos(kL) - 2t_{\text{hop},2}\cos(2kL). \qquad (\text{C.8})$$

The effect on $E(k)$ of including $t_{\text{hop},2} = 0.2 \times t_{\text{hop},1}$ and $t_{\text{hop},2} = 0.5 \times t_{\text{hop},1}$ is illustrated in figure C.1(a) and (b) respectively. Incorporation of next-nearest-neighbor

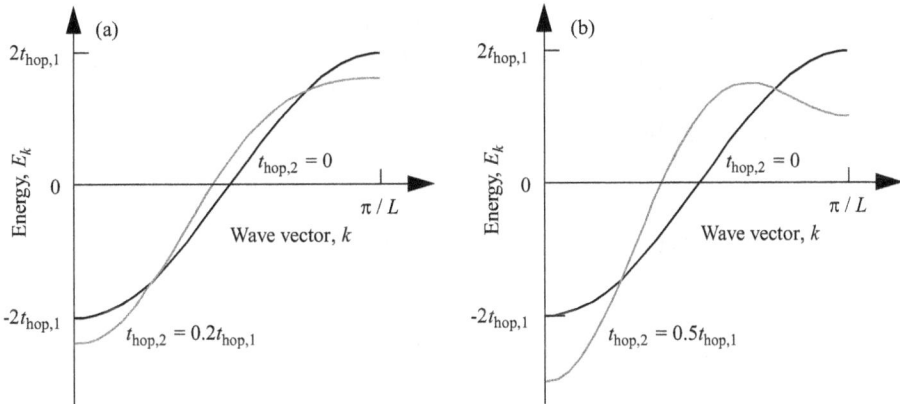

Figure C.1. (a) Electron energy dispersion relation E_k for a one-dimensional crystal with nearest-neighbor atom spacing L and s-atomic orbitals at lattice sites. Energy E_k is calculated in the tight-binding approximation for nearest-neighbor interaction ($t_{\text{hop},1} \neq 0$) giving a cosine band of energy width $E_b = 4 \times t_{\text{hop},1}$ (black curve). Inclusion of next-nearest-neighbor hopping with $t_{\text{hop},2} = 0.2 \times t_{\text{hop},1}$ modifies the electron dispersion relation, E_k (red curve). (b) Same as (a) but with next-nearest-neighbor hopping such that $t_{\text{hop},2} = 0.5 \times t_{\text{hop},1}$ (red curve).

terms increases the complexity of the dispersion relation allowing, for example, different curvature at band extremes. The next-nearest-neighbor terms add a Fourier component to $E(k)$, providing access to dispersion not available in the Kronig–Penney model described in section 1.5.

The dispersion relation for nearest-neighbor hopping given by equation (C.7) in a cubic lattice with $L = L_x = L_y = L_z$ is

$$E_k = -2t_{\text{hop},1} \left(\cos(k_x L) + \cos(k_y L) + \cos(k_z L) \right). \tag{C.9}$$

In this case, the three-dimensional cosine tight-binding band for propagating electron states has an energy bandwidth of $E_b = 12 \times t_{\text{hop},1}$.

C.2 Tight-binding complex band structure

The geometry and single crystal heterostructures used to create semiconductor laser devices have electronic states that are not those described by bulk band structure. For example, the electronic states of a semiconductor quantum well depend on the thickness of the potential well. Of course, localized single-electronic states in a quantum well, at a surface, interface, heterojunction, and defect, are found by solving the Schrödinger equation for the specific configuration under consideration. This can be quite involved since, for each energy an electron has at an interface, all the solutions of the Schrödinger equation, including the complex states, must be known [1]. However, in practice, it is often the case that intuition and heuristics about these solutions can be inferred from bulk band structure and, in particular for localized states, from complex band structure. It is in this way that the tight-binding complex band structure of bulk crystals [2, 3] can act as a guide to finding a suitable starting point for the optimal design of nanoscale semiconductor devices, including the determination of optimal current–voltage characteristics that are controlled by electron scattering [4, 5].

The dispersion relation for electron states in a one-dimensional tight-binding model describing s-orbital atoms with constant lattice constant L and nearest and next-nearest-neighbor hopping may be written as equation (C.8), $E_k = -2t_{\text{hop},1} \cos(kL) -2t_{\text{hop},2} \cos(2kL)$. Using the relations $2 \sin(x) \sin(y) = \cos(x - y) - \cos(x + y)$ and $\sin^2(x) = 1 - \cos^2(x)$ with $x = y = kL$ to give $\cos(2kL) = 2 \cos^2(kL) - 1$, then

$$2t_{\text{hop},1} \cos(kL) + 2t_{\text{hop},2} \left(2 \cos^2(kL) - 1 \right) - E = 0, \tag{C.10}$$

which is a quadratic function

$$4t_{\text{hop},2} \cos^2(kL) + 2t_{\text{hop},1} \cos(kL) - 2t_{\text{hop},2} - E = 0 \tag{C.11}$$

of the form $a\theta_B^2 + b\theta_B + c = 0$ with solutions $\theta_B = \frac{-b \pm \sqrt{b^2 - 4ac}}{2a}$ and, since $\theta_B = \cos(kL)$, the solutions for delocalized propagating electron states with normalized kL at a given energy E are equation (1.43), $kL = \frac{\text{acos}(\theta_B)}{\pi}$ with $|\theta_B| < 1$, and for localized non-propagating electron states equation (1.44), $kL = \frac{\text{acosh}(\theta_B)}{\pi}$ with

$|\theta_B| > 1$. A delocalized non-propagating state with real wave vector exists when $|\theta_B| = 1$.

In general, kL is complex with two branches arising from the solution to the quadratic equation due to the presence of next-nearest-neighbor hopping. Pure real kL describes delocalized propagating states of the form e^{ikx} and complex $kL = \mathrm{Re}\,(kL) + \mathrm{i}\,\mathrm{Im}(kL)$ describes localized non-propagating states with an exponentially decaying envelope of the form $e^{\mp k_{\mathrm{Im}}x}$, where $k_{\mathrm{Im}} = \mathrm{Im}(k) \neq 0$. These evanescent electronic states grow or decay exponentially from one unit cell to the next, are unbounded (diverge) at $x \to \mp\infty$, and are excluded by translational symmetry in a bulk crystal. However, complex band structures can be useful as a guide and an efficient way to estimate exponential spatial decay of electronic states that are the solution to the Schrödinger equation in finite-sized semiconductor structures, heterointerfaces, tunnel barriers, surfaces, and defects in nanoscale devices.

Figure C.2 illustrates the complex band structure in a one-dimensional crystal with nearest-neighbor atom spacing L and an s-atomic orbital at each lattice site when next-nearest-neighbor hopping energy $t_{\mathrm{hop},2} = 0.5 \times t_{\mathrm{hop},1}$. Notice that complex values of k in the complex plane analytically connect to the pure real values of k in the real plane at a point when $|\theta_B| = 1$ and where dispersion in the real and complex plane has zero derivatives, $\mathrm{d}E/\mathrm{d}k = 0$. This point is singular (a Van Hove singularity) in the density of propagating states. A band extremum exists when $\mathrm{d}E/\mathrm{d}k = 0$, electron group velocity is zero, and the corresponding electron state that connects real and complex values of wave vector has a real wave vector describing a state that is delocalized and non-propagating.

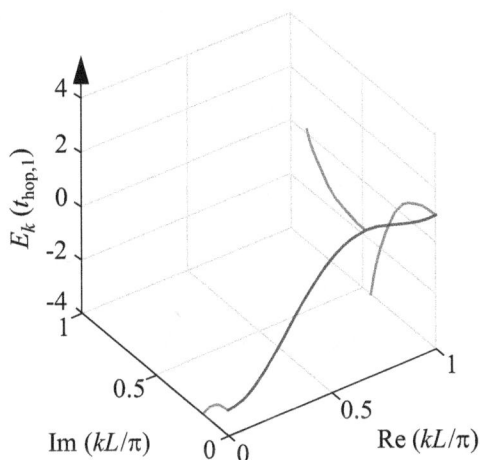

Figure C.2. Complex dispersion relation as a function of electron energy, E_k, in a one-dimensional crystal with nearest-neighbor atom spacing L and an s-atomic orbital at each lattice site. In this example, the dispersion relation is calculated in the tight-binding approximation for nearest-neighbor hopping energy $t_{\mathrm{hop},1}$ and next-nearest-neighbor hopping energy $t_{\mathrm{hop},2} = 0.5 \times t_{\mathrm{hop},1}$. Propagating electron states with pure real wave vectors are indicated by the blue curve. States that do not propagate in the bulk have complex wave vectors and are indicated by the red curves.

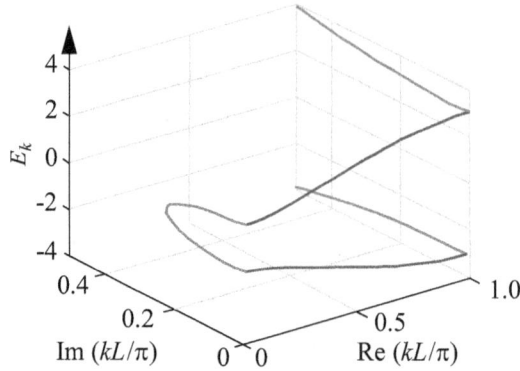

Figure C.3. Complex dispersion relation as a function of electron energy, E_k, in a one-dimensional crystal with nearest-neighbor atom spacing L and alternating normalized nearest-neighbor hopping energy $t_{hop,a} = 1$ and $t_{hop,b} = 2$. Propagating electron states with pure real wave vectors are indicated by the blue curves. States that do not propagate in the bulk have complex wave vectors and are indicated by the red curves. The band gap energy separating the valence band and conduction band is $E_g = 2\,|t_{hop,a} - t_{hop,b}|$.

Because a tight-binding complex band structure results from an expansion in localized atomic orbitals with nearest and next-nearest-neighbor hopping energy, it contains more features than the dispersion of an electron subject to a simple periodic array of rectangular potential barriers previously described in section 1.5.

To illustrate a complex band structure with a valence and conduction band separated by a band gap of energy E_g, consider a one-dimensional lattice of atoms with nearest-neighbor spacing L and nearest-neighbor tight-binding interaction energy $t_{hop,a}$ and $t_{hop,b}$ that alternates between atom sites. For zero onsite potential, the Hamiltonian is

$$\hat{H}(k) = \begin{pmatrix} 0 & \alpha \\ \alpha^* & 0 \end{pmatrix}, \tag{C.12}$$

where $\alpha = t_{hop,a} + t_{hop,b}e^{ikL}$. The corresponding Schrödinger equation has energy eigenvalues

$$E(k) = \pm|\alpha| = \pm\sqrt{t_{hop,a}^2 + t_{hop,b}^2 + 2t_{hop,a}t_{hop,b}\cos(kL)} \tag{C.13}$$

and complex band structure shown in figure C.3.

References

[1] Heine V 1963 *Proc. Phys. Soc. London* **81** 300
[2] Chang Y-C 1982 *Phys. Rev.* B **25** 605
[3] Reuter M G 2017 *J. Phys. Condens. Matter* **29** 053001
[4] Unglaub W and Levi A F J 2023 *Phys. Open* **17** 100164
[5] Unglaub W and Levi A F J 2025. *Physica* E **165** 116067

Essential Semiconductor Laser Device Physics (Second Edition)

A F J Levi

Appendix D

The beam splitter

D.1 Transmission and reflection of a single photon at a beam splitter

Consider the ideal, lossless, symmetric, 50:50 beam splitter illustrated in figure D.1. It is configured with two input ports and two output ports. The quantum field reflection and transmission coefficients for a single photon entering port 1 or 2 are $r_{\text{ph},1}$, $r_{\text{ph},2}$, $t_{\text{ph},1}$, and $t_{\text{ph},2}$, respectively. If a single isolated photon enters port 1 then it is in the input state $| n_1 = 1, \ n_2 = 0 \rangle_{\text{in}}$. For the case being considered, it is well known that the phase of the transmitted field leads the phase of the reflected field by $\pi/2$ [1]. To see why is considered next.

The single-photon input state $|n_1 = 1, \ n_2 = 0 \rangle_{\text{in}}$ has quantum field amplitude at port 1 that can be set to $a_1 = 1$ and at port 2 it is set to $a_1 = 0$. In this case the input state is a product state so that $|1, \ 0 \rangle_{\text{in}} = |1 \rangle_1 \otimes |0 \rangle_2$, where $|0 \rangle$ is the vacuum state. For input state $|n_1 = 0, \ n_2 = 1 \rangle_{\text{in}}$, the quantum field output is $a_3 = r_{\text{ph},2}$ at port 3 and $a_4 = t_{\text{ph},2}$ at port 4. The single-photon input and output *quantum field amplitudes* are related via

$$\begin{bmatrix} a_3 \\ a_4 \end{bmatrix}_{\text{out}} = \begin{bmatrix} t_{\text{ph},1} & r_{\text{ph},2} \\ r_{\text{ph},1} & t_{\text{ph},2} \end{bmatrix} \begin{bmatrix} a_1 \\ a_2 \end{bmatrix}_{\text{in}} = \hat{U}_{\text{B}} \begin{bmatrix} a_1 \\ a_2 \end{bmatrix}_{\text{in}}, \quad (\text{D.1})$$

where $\hat{U}_{\text{B}}$ is a 2×2 matrix describing the ideal lossless beam splitter.

Photon probability is conserved because an ideal lossless beam splitter is being considered. This means that the 2×2 matrix $\hat{U}_{\text{B}}$ is unitary and, consequently, its Hermitian adjoint is its inverse, $\hat{U}_B^\dagger = \hat{U}_B^{-1}$. Hence,

$$\hat{U}_B^\dagger = \begin{bmatrix} t_{\text{ph},1}^* & r_{\text{ph},1}^* \\ r_{\text{ph},2}^* & t_{\text{ph},2}^* \end{bmatrix} = \frac{1}{t_{\text{ph},1}t_{\text{ph},2} - r_{\text{ph},1}r_{\text{ph},2}} \begin{bmatrix} t_{\text{ph},2} & -r_{\text{ph},2} \\ -r_{\text{ph},1} & t_{\text{ph},1} \end{bmatrix} = \hat{U}_B^{-1}. \quad (\text{D.2})$$

doi:10.1088/978-0-7503-6417-1ch11

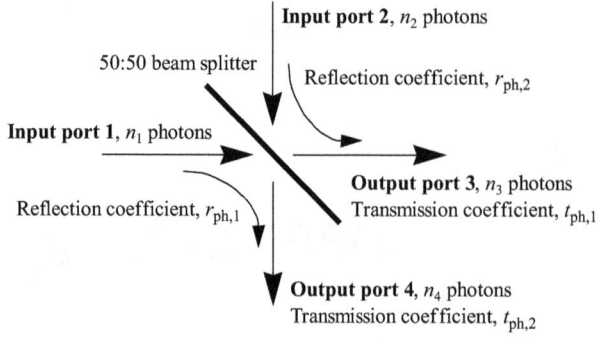

Figure D.1. Illustration of a lossless 50:50 beam splitter showing n_1 photons incident at input port 1, n_2 photons incident at input port 2, n_3 photons emerge from output port 3, and n_4 photons emerge from output port 4. Reflection from the lossless symmetric 50:50 beam splitter introduces a $\pi/2$ phase shift relative to the transmitted beam.

Because the determinant of a unitary matrix has unit magnitude, in general,

$$t_{\text{ph},1}t_{\text{ph},2} - r_{\text{ph},1}r_{\text{ph},2} = e^{i\varphi}, \tag{D.3}$$

where φ is a global phase factor that has no impact on the relative phase between matrix elements and may be set to $\varphi = \pi$, so that

$$t_{\text{ph},1}t_{\text{ph},2} - r_{\text{ph},1}r_{\text{ph},2} = -1 \tag{D.4}$$

since $e^{i\pi} = -1$. Inserting this into the expression for $\hat{U}_B^\dagger$ gives

$$r_{\text{ph},1} = r_{\text{ph},2}^* \tag{D.5}$$

and

$$t_{\text{ph},1} = -t_{\text{ph},2}^* \tag{D.6}$$

from which it may be concluded that $|r_{\text{ph},1}| = |r_{\text{ph},2}|$ and $|t_{\text{ph},1}| = |t_{\text{ph},2}|$. Re-expressing the complex terms for $r_{\text{ph},1}$ and $t_{\text{ph},1}$ gives

$$|r_{\text{ph},1}|e^{i\theta_{r_{\text{ph},1}}} = |r_{\text{ph},2}|e^{-i\theta_{r_{\text{ph},2}}} \tag{D.7}$$

and

$$|t_{\text{ph},1}|e^{i\theta_{t_{\text{ph},1}}} = -|t_{\text{ph},2}|e^{-i\theta_{t_{\text{ph},2}}}. \tag{D.8}$$

Dividing these equations gives

$$\frac{|t_{\text{ph},1}|e^{i\theta_{t_{\text{ph},1}}}}{|r_{\text{ph},1}|e^{i\theta_{r_{\text{ph},1}}}} = \frac{-|t_{\text{ph},2}|e^{-i\theta_{t_{\text{ph},2}}}}{|r_{\text{ph},2}|e^{-i\theta_{r_{\text{ph},2}}}} \rightarrow e^{i\theta_{t_{\text{ph},1}} - i\theta_{r_{\text{ph},1}}} = -e^{-i\theta_{t_{\text{ph},2}} + i\theta_{r_{\text{ph},2}}} = e^{-i\theta_{t_{\text{ph},2}} + i\theta_{r_{\text{ph},2}} + i\pi}, \tag{D.9}$$

where use is made of $e^{i\pi} = -1$. Hence,

$$(\theta_{t_{ph,1}} - \theta_{r_{ph,1}}) + (\theta_{t_{ph,2}} - \theta_{r_{ph,2}}) = \pi. \tag{D.10}$$

For the ideal, lossless, symmetric, 50:50 beam splitter $r_{ph,1} = r_{ph,2}$ and $t_{ph,1} = t_{ph,2}$. Therefore, the phase difference between transmission and reflection at each port is the same,

$$(\theta_{t_{ph,1}} - \theta_{r_{ph,1}}) = (\theta_{t_{ph,2}} - \theta_{r_{ph,2}}) = \frac{\pi}{2} \tag{D.11}$$

and it is clear that the phase of the transmitted field leads the phase of the reflected field by $\pi/2$. For the perfect, lossless, symmetric, 50:50 dielectric beam splitter $|r_{ph,1}| = |r_{ph,2}| = |t_{ph,1}| = |t_{ph,2}|$, which can only be satisfied if $r_{ph,1} = r_{ph,2} = r_{ph}$ is real and $t_{ph,1} = t_{ph,2} = t_{ph}$ is pure imaginary. Given the fact that the determinant of the unitary matrix requires $t_{ph,1}t_{ph,2} - r_{ph,1}r_{ph,2} = -1$, then

$$r_{ph} = \frac{-1}{\sqrt{2}} \tag{D.12}$$

and

$$t_{ph} = \frac{i}{\sqrt{2}} \tag{D.13}$$

so that the unitary 2×2 matrix describing a single photon interacting with an ideal, lossless, symmetric, 50:50 beam splitter is

$$\hat{U}_B = \frac{1}{\sqrt{2}} \begin{bmatrix} i & -1 \\ -1 & i \end{bmatrix} = \begin{bmatrix} t_{ph} & r_{ph} \\ r_{ph} & t_{ph} \end{bmatrix}, \tag{D.14}$$

which satisfies the unitary requirement $\hat{U}_B^\dagger = \hat{U}_B^{-1}$.

D.2 An integer number of photons at each input port of a beam splitter

A photon-number input state $|n_1, n_2\rangle_{in}$ corresponds to *integer* n_1 photons incident at port 1 and *integer* n_2 photons incident at port 2. The output of the beam splitter has photons with quantum field amplitude a_3 at port 3 and amplitude a_4 at port 4. The input state of the beam splitter $|n_1, n_2\rangle_{in}$ is connected to the output state of the beam splitter $|a_3, a_4\rangle_{out}$ by a linear transformation $|a_3, a_4\rangle_{out} = \mathbf{B} |n_1, n_2\rangle_{in}$ where $\mathbf{B}$ is a 2×2 matrix

$$\mathbf{B} = \begin{bmatrix} t_{ph,1} & r_{ph,2} \\ r_{ph,1} & t_{ph,2} \end{bmatrix} \tag{D.15}$$

so that

$$\begin{bmatrix} a_3 \\ a_4 \end{bmatrix} = \mathbf{B} \begin{bmatrix} n_1 \\ n_2 \end{bmatrix} = \begin{bmatrix} t_{ph,1} & r_{ph,2} \\ r_{ph,1} & t_{ph,2} \end{bmatrix} \begin{bmatrix} n_1 \\ n_2 \end{bmatrix}. \tag{D.16}$$

If the photons incident on the beam splitter are *indistinguishable* then the probability of transmission or reflection must take into account the number of ways of arranging the photons among themselves. For example, with $n_2 = 0$, the probability that a single beam of n_1 indistinguishable photons are transmitted as n_3 photons and reflected as $n_1 - n_3 = n_4$ photons at an ideal, lossless, symmetric, 50:50 beam splitter is given by

$$P(n_1, n_2 = 0, \ n_3, \ n_4 = n_1 - n_3) = \frac{n_1!}{n_3!(n_1 - n_3)!}\left(\frac{1}{2}\right)^{n_1} = \binom{n_1}{n_3}\left(\frac{1}{2}\right)^{n_1}, \quad (D.17)$$

where

$$\binom{n_1}{n_3} = \frac{n_1!}{n_3!(n_1 - n_3)!} \quad (D.18)$$

is the binomial coefficient.

If the photons are *distinguishable*, the probability that a single beam of n_1 photons are transmitted as n_3 photons at the beam splitter would have the smaller value

$$P(n_1, \ 0, \ n_3, \ n_4) = \left(\frac{1}{2}\right)^{n_1}. \quad (D.19)$$

Photons are indistinguishable if they have the same polarization, the same frequency, the same phase, and arrive at the detector at the same time. Photon polarization, frequency, and phase are internal degrees of freedom. The direction of motion is not used to distinguish photons. One way to continuously tune the system from quantum (indistinguishable) to classical (distinguishable) behavior is to introduce a delay between the detected time of arrival of photons [2].

The detected output-state probabilities of a single photon interacting with an ideal, lossless, symmetric, 50:50 beam splitter are the same as expected for a classical electromagnetic wave. However, if *two* identical indistinguishable photons interact with the beam splitter then dramatically different predictions are obtained that are contrary to the expectations of unmodulated classical light.

D.3 The Mandel effect: transmission of two indistinguishable photons at a beam splitter

In general, if there is an integer number of photons at the inputs of a beam splitter then the Fock state is $|n_1, \ n_2\rangle_{in}$ with integer n_1 photons at input port 1 and integer n_2 photons at input port 2. Likewise, a Fock output state of the beam splitter, $|n_3, \ n_4\rangle_{out}$, has n_3 photons at output port 3 and n_4 photons at output port 4.

A bi-photon source can be used to create two indistinguishable photons. These can be input to the two input ports of an ideal, lossless, symmetric 50:50 beam splitter. Every possible combination of inferred photon paths through the beam splitter that is consistent with the exchange-symmetric product states for the boson two-particle system must be accounted for. In contrast to the description of classical *particles*, indistinguishable quantum particles may be viewed as *simultaneously* experiencing *every possible path* through the system.

If one photon is introduced at input port 1 and the other at input port 2 then the input state is $|n_1 = 1,\ n_2 = 1\rangle_{\text{in}}$. This input state is transformed to output states by passing through the beam splitter so that $|n_1 = 1,\ n_2 = 1\rangle_{\text{in}} \rightarrow |n_3,\ n_4\rangle_{\text{out}}$. When $n_1 = n_2 = 1$ there are just three exchange-symmetric output product states. Setting reflection coefficient $r_{\text{ph}} = -1$ and transmission coefficient $t_{\text{ph}} = i$ results in non-normalized output states

$$|n_1 = 1,\ n_2 = 1\rangle_{\text{in}} \rightarrow |n_3 = 2,\ n_4 = 0\rangle_{\text{out}} : \ t_{\text{ph}} r_{\text{ph}} = -i \tag{D.20}$$

$$|n_1 = 1,\ n_2 = 1\rangle_{\text{in}} \rightarrow |n_3 = 1,\ n_4 = 1\rangle_{\text{out}} : \ \frac{r_{\text{ph}} r_{\text{ph}} + t_{\text{ph}} t_{\text{ph}}}{\sqrt{2}} = \frac{(1-1)}{\sqrt{2}} = 0 \tag{D.21}$$

$$|n_1 = 1,\ n_2 = 1\rangle_{\text{in}} \rightarrow |n_3 = 0,\ n_4 = 2\rangle_{\text{out}} : \ r_{\text{ph}} t_{\text{ph}} = -i. \tag{D.22}$$

For the case $|n_1 = 1,\ n_2 = 1\rangle_{\text{in}} \rightarrow |n_3 = 1,\ n_4 = 1\rangle_{\text{out}}$ there are two indistinguishable paths the photons can take from input to output that can be inferred after detection. They are either both reflected or both transmitted through the beam splitter. Each inferred path after detection is equally likely in the 50:50 beam splitter, so the photons may be considered to be simultaneously experiencing both processes with the same weight. The superposition of product amplitudes $(r_{\text{ph}} r_{\text{ph}} + t_{\text{ph}} t_{\text{ph}})/\sqrt{2}$ describes this. The photon field amplitude interference that results in $(r_{\text{ph}} r_{\text{ph}} + t_{\text{ph}} t_{\text{ph}})/\sqrt{2} = 0$ occurs because the detectors are unable to distinguish between the two two-photon paths.

The photon-number detection probabilities at the output ports are *proportional* to the absolute value squared of the non-normalized quantum amplitudes

$$P_{\text{out}}^{\text{non}}(n_1 = 1,\ n_2 = 1,\ n_3 = 2,\ n_4 = 0) = |t_{\text{ph}} r_{\text{ph}}|^2 = 1 \tag{D.23}$$

$$P_{\text{out}}^{\text{non}}(n_1 = 1,\ n_2 = 1,\ n_3 = 1,\ n_4 = 1) = \left|\frac{r_{\text{ph}} r_{\text{ph}} + t_{\text{ph}} t_{\text{ph}}}{\sqrt{2}}\right|^2 = \frac{|1-1|^2}{2} = 0 \tag{D.24}$$

$$P_{\text{out}}^{\text{non}}(n_1 = 1,\ n_2 = 1,\ n_3 = 0,\ n_4 = 2) = |r_{\text{ph}} t_{\text{ph}}|^2 = 1. \tag{D.25}$$

Normalization of $P_{\text{out}}^{\text{non}}$ output values in equations (D.23)–(D.25) enable interpretation as probability, P_{out}. Normalization may be achieved via division by the sum

$$P_{\text{sum}} = \sum_j P_{j,\text{out}}^{\text{non}} \tag{D.26}$$

In this case $P_{\text{sum}} = 2$ and so the normalized probability values are

$$P_{\text{out}}(n_1 = 1,\ n_2 = 1,\ n_3 = 2,\ n_4 = 0) = \frac{|t_{\text{ph}} r_{\text{ph}}|^2}{P_{\text{sum}}} = \frac{1}{2} \tag{D.27}$$

$$P_{\text{out}}(n_1 = 1,\ n_2 = 1,\ n_3 = 1,\ n_4 = 1) = 0 \tag{D.28}$$

$$P_{\text{out}}(n_1 = 1,\ n_2 = 1,\ n_3 = 0,\ n_4 = 2) = \frac{|r_{\text{ph}} t_{\text{ph}}|^2}{P_{\text{sum}}} = \frac{1}{2}. \tag{D.29}$$

If there is one indistinguishable photon at each input port, then *the detected quantum amplitudes interfere and cancel exactly*, so there is precisely *zero* probability of detecting one photon at each output port. The zero probability of detecting a $|n_3 = 1,\ n_4 = 1\rangle_{\text{out}}$ output state when there is a $|n_1 = 1,\ n_2 = 1\rangle_{\text{in}}$ input state is a strong quantum correlation effect first measured by Hong, Ou, and Mandel [3]. In addition, one indistinguishable photon at each input port can *only* result in *two* photons detected at an output port. This effect, also driven by the symmetry of identical, indistinguishable, boson particles in quantum mechanics, is an example of *photon bunching*.

If two photons are introduced at input port 1 and zero at input port 2, then the input state is $|n_1 = 2,\ n_2 = 0\rangle_{\text{in}}$ and the possible output-state amplitudes are proportional to

$$|n_1 = 2,\ n_2 = 0\rangle_{\text{in}} \rightarrow |n_3 = 2,\ n_4 = 0\rangle_{\text{out}} :\ t_{\text{ph}} t_{\text{ph}} = -1 \tag{D.30}$$

$$|n_1 = 2,\ n_2 = 0\rangle_{\text{in}} \rightarrow |n_3 = 1,\ n_4 = 1\rangle_{\text{out}} :\ \frac{t_{\text{ph}} r_{\text{ph}} + r_{\text{ph}} t_{\text{ph}}}{\sqrt{2}} = \frac{-2i}{\sqrt{2}} \tag{D.31}$$

$$|n_1 = 2,\ n_2 = 0\rangle_{\text{in}} \rightarrow |n_3 = 0,\ n_4 = 2\rangle_{\text{out}} :\ r_{\text{ph}} r_{\text{ph}} = 1. \tag{D.32}$$

In this case $P_{\text{sum}} = 4$ and the corresponding photon-number detection probabilities at the output ports are

$$P_{\text{out}}(n_1 = 2,\ n_2 = 0,\ n_3 = 2,\ n_4 = 0) = \frac{|t_{\text{ph}} t_{\text{ph}}|^2}{P_{\text{sum}}} = \frac{1}{4} \tag{D.33}$$

$$P_{\text{out}}(n_1 = 2,\ n_2 = 0,\ n_3 = 1,\ n_4 = 1) = \frac{|t_{\text{ph}} r_{\text{ph}} + r_{\text{ph}} t_{\text{ph}}|^2}{2P_{\text{sum}}} = \frac{1}{2} \tag{D.34}$$

$$P_{\text{out}}(n_1 = 2,\ n_2 = 0,\ n_3 = 0,\ n_4 = 2) = \frac{|r_{\text{ph}} r_{\text{ph}}|^2}{P_{\text{sum}}} = \frac{1}{4}. \tag{D.35}$$

When the total number of indistinguishable photons $n_{\text{tot}} = 2$ there are three possible input states and three possible output states with detected output probabilities as indicated in table D.1 and represented graphically in figure D.2. If there is one indistinguishable photon at each input port then the detected output is two photons at one of the output ports. In general, if there is an equal number of indistinguishable photons at each input port ($n_1 = n_2$) then it is not possible to have an odd number of photons at an output port.

Two-photon quantum interference (the $n_{\text{tot}} = 2$ case) is not interference of two separate photons at the beam splitter, but rather it is interference of the two two-photon amplitudes at the detectors. The photon paths can only be inferred *after* detection at the detectors. This is a consequence of the standard Copenhagen

Table D.1. Output probabilities of two identical, indistinguishable, photons interacting with an ideal, lossless, symmetric, 50:50 beam splitter.

Input state $|n_1,\ n_2\rangle$	Output state $|n_3,\ n_4\rangle$		
	$|2,\ 0\rangle$	$|1,\ 1\rangle$	$|0,\ 2\rangle$
$|2,\ 0\rangle$	$\frac{1}{4}$	$\frac{1}{2}$	$\frac{1}{4}$
$|1,\ 1\rangle$	$\frac{1}{2}$	0	$\frac{1}{2}$
$|0,\ 2\rangle$	$\frac{1}{4}$	$\frac{1}{2}$	$\frac{1}{4}$

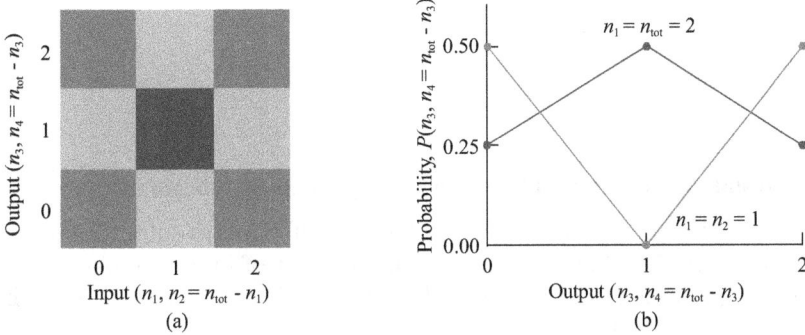

Figure D.2. (a) Probability of photon-number detection as a function of the input and output states of an ideal, lossless, symmetric, 50:50 beam splitter. There are n_1 photons at input port 1 and n_2 photons at input port 2 when there is a total of $n_{tot} = 2$ indistinguishable photons in the system. For the case when $n_1 = 1$ and $n_2 = 1$ there is zero probability that the value of $n_3 = 1$ and that the value of $n_4 = 1$. (b) The dots show the output probability into output port 3 for the case when the input port 1 contains $n_1 = n_{tot} = 2$ and when $n_1 = n_2 = n_{tot}/2 = 1$. The lines connecting the dots are to guide the eye.

interpretation of quantum mechanics in which the properties of a system are only obtained *after* interaction between the quantum system and the measurement instrument—in this case, the detectors. In fact, photons do not have to arrive simultaneously at the beam splitter to have their quantum field amplitudes interfere at the detectors; rather it is only that the inferred two two-photon paths must be indistinguishable [4]. In practice, it is the *inferred* interpretation of indistinguishable two-photon paths that may be used as a convenient way to predict quantum field amplitude interference.

D.4 Transmission of *n* indistinguishable photons at a beam splitter

The quantum amplitude of integer n_3 and n_4 indistinguishable photons appearing at the output ports of the beam splitter is

$$|n_1,\ n_2,\ n_3,\ n_4\rangle = (-1)^{n_1}\left(\frac{1}{2}\right)^{\frac{n_1 + n_2}{2}}\sum_j(-1)^j\sqrt{\binom{n_1}{j}\binom{n_2}{n_3 - j}\binom{n_3}{j}\binom{n_4}{n_1 - j}}, \quad (D.36)$$

where, because the total number of particles, $n_{tot} = n_1 + n_2$, is conserved, $n_4 = n_1 + n_2 - n_3$. In this expression, the number of ways of choosing j photons from a set of n_{tot} photons is given by the binomial coefficient

$$\binom{n_{tot}}{j} = \frac{n_{tot}!}{n_{tot}!(n_{tot} - j)!} \qquad (D.37)$$

If j is negative or greater than n_{tot} the binomial coefficient is set to zero. The terms $(-1)^{n_1}$ and $(-1)^j$ are due to the relative phase difference between a transmitted or reflected photon and it is this that gives rise to strong quantum interference effects. When using equation (D.37), the probability of detecting photons at the output ports of the beam splitter is

$$P_{out}(n_1, \ n_2, \ n_3, \ n_4) = ||n_1, \ n_2, \ n_3, \ n_4\rangle|^2 . \qquad (D.38)$$

D.4.1 Transmission of eight indistinguishable photons at a beam splitter

Figure D.3(a) shows the calculated probability of photon output from an ideal, lossless symmetric 50:50 beam splitter with the input of integer n_1 photons at input port 1 and n_2 photons at input port 2 when there is a total of $n_{tot} = 8$ indistinguishable photons in the system.

The blue dots in figure D.3(b) are port 3 output probability for the case when the input port 1 contains $n_1 = n_{tot}$ indistinguishable photons. The probability has an approximately normal distribution centered at $n_{tot}/2$. The red dots show port 3 output probability for the case when the input port 1 contains $n_1 = n_2 = n_{tot}/2 = 4$

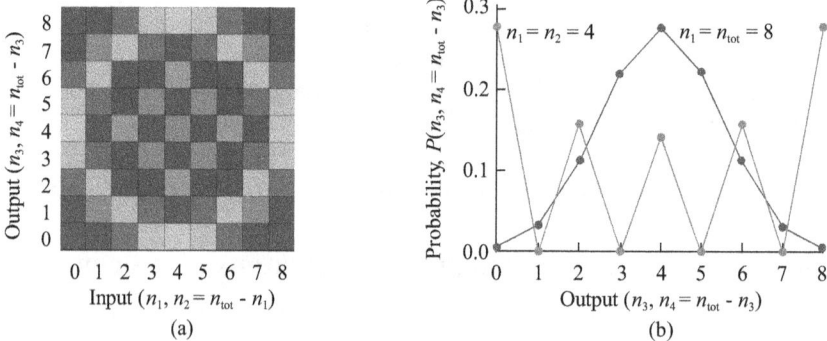

Figure D.3. (a) Probability of photon output from a lossless symmetric 50:50 beam splitter showing input of n_1 photons at input port 1 and n_2 photons at input port 2 when there is a total of $n_{tot} = 8$ indistinguishable photons in the system. (b) The dots shows output probability into output port 3 for the case when the input port 1 contains $n_1 = n_{tot} = 8$ and when $n_1 = n_2 = n_{tot}/2 = 4$. The lines connecting the dots are to guide the eye.

indistinguishable photons. In this situation, symmetry dictates that the probability of an odd number of indistinguishable photons at an output port is zero. The system behavior is closest to the expectations of a continuous unmodulated classical electromagnetic wave when photons are only present at one input port. The system behavior is most non-classical when there are equal numbers of photons at each input port of the beam splitter. Increasing the value of n_{tot} can be used to explore the transition between the most non-classical behavior and behavior that seems closest to classical expectations. As will be illustrated next, near-classical results for a continuous unmodulated electromagnetic wave are retrieved when either $n_1 = n_{tot}$ or $n_2 = n_{tot}$ in the limit $n_{tot} \rightarrow \infty$.

D.4.2 Transmission of 64 indistinguishable photons at a beam splitter

Figure D.4 shows the results of calculating the transmission of $n_{tot} = 64$ identical, indistinguishable photons at an ideal, lossless, symmetric, 50:50 beam splitter. The blue curve in figure D.4(b) is port 3 detected output probability for the case when the input port 1 contains $n_1 = n_{tot}$ indistinguishable photons. The probability has an approximately normal distribution centered at $n_{tot}/2$ and full-width-half-maximum (FWHM) slightly greater than $8 = \sqrt{64}$. In the limit when $n_{tot} \rightarrow \infty$ (the large particle number thermodynamic limit) and $n_1 = n_{tot}$ the detected photon-number output probability distribution exhibits the normal (classical) result FWHM $\rightarrow \sqrt{n_{tot}}$. The red curve in figure D.4(b) is port 3 detected output probability showing the most non-classical behavior. This occurs when $n_1 = n_2 = n_{tot}/2 = 32$.

Figure D.5 is a three-dimensional plot of the probability of photon output from an ideal, lossless, symmetric, 50:50 beam splitter when there is a total of $n_{tot} = 64$ indistinguishable photons in the system. The appearance of detected quantum

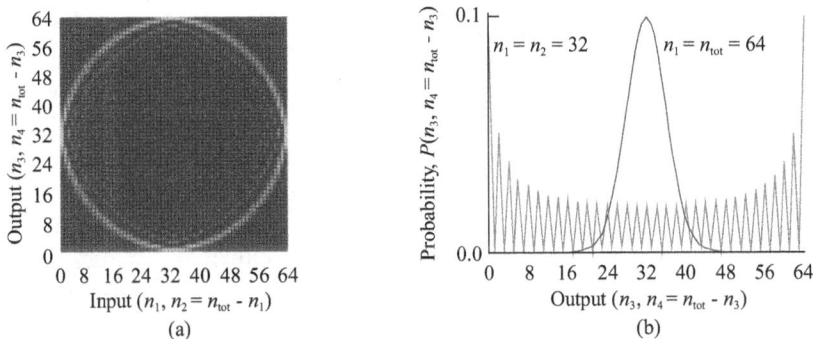

Figure D.4. (a) Probability of detected photon output from an ideal, lossless, symmetric, 50:50 beam splitter showing input of n_1 photons at input port 1 and n_2 photons at input port 2 when there is a total of $n_{tot} = 64$ indistinguishable photons in the system. (b) Port 3 detected output probability for the case when the input port 1 contains $n_1 = n_{tot}$ and when $n_1 = n_2 = n_{tot}/2 = 32$. The lines connecting probability values on the vertical axis for integer values of n_3 on the horizontal axis are to guide the eye.

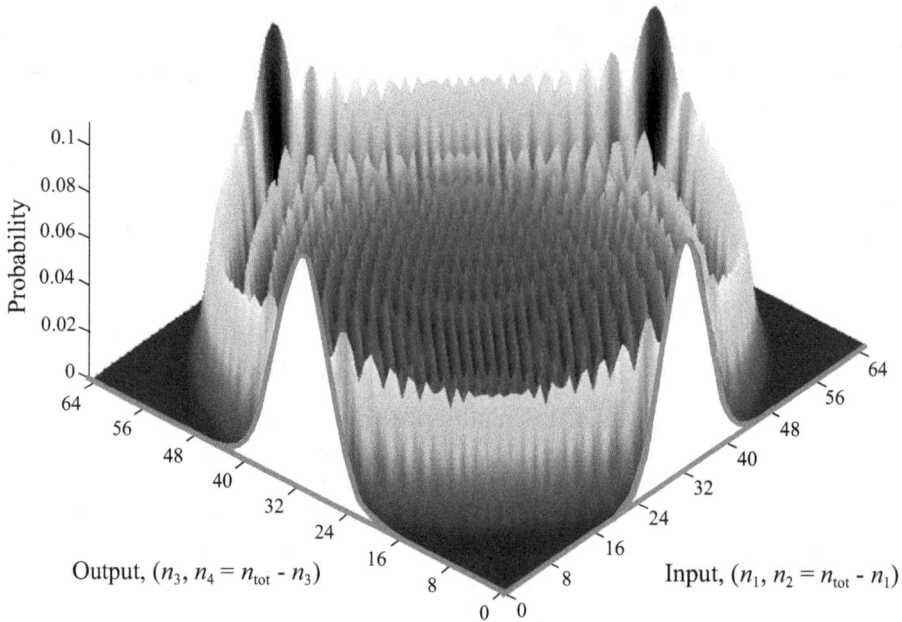

Figure D.5. Three-dimensional plot of probability of detected photon output from an ideal, lossless, symmetric, 50:50 beam splitter when there is a total of $n_{\text{tot}} = 64$ indistinguishable photons in the system. There is a 'quantum cauldron' of interference inside a radius $n_{\text{tot}}/2$. The closest to classical behavior occurs at the boundary of the domain when $n_1 = n_{\text{tot}}$ and $n_2 = 0$ or when $n_1 = 0$ and $n_2 = n_{\text{tot}}$.

interference effects is limited to a 'quantum cauldron' of radius $n_{\text{tot}}/2$ [5]. The closest to classical behavior (an approximately normal probability distribution) occurs at the boundary of the domain when $n_1 = n_{\text{tot}}$ and $n_2 = 0$ or $n_1 = 0$ and $n_2 = n_{\text{tot}}$. The quantum cauldron illustrated in figure D.5 is one way to depict the transition from closest to classical behavior to most non-classical behavior in the system.

In the preceding, the photon number is preserved, and the interaction of the optical field with the beam splitter is ideal. The symmetry associated with indistinguishable particles results in detected quantum amplitude interference between different inferred paths through the system. Again, it should be noted that fundamental to the Copenhagen interpretation of quantum mechanics, the photon paths taken can only be inferred *after* detection. As illustrated by the Mandel effect [3], because of strong quantum correlations, the probability of detecting a fixed number of discrete photons at an output port can be dramatically different from the expectations of a continuous unmodulated classical electromagnetic wave interacting with the system.

D.5 Quantum interference and distinguishability

The Mandel effect [3] is an example of quantum interference whose effect is maximized if the photon particles in the system are indistinguishable.

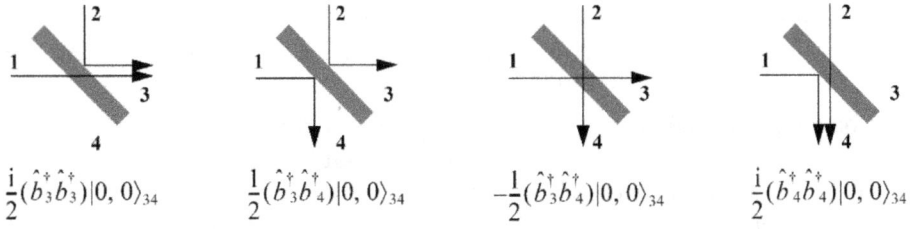

$$\frac{i}{2}(\hat{b}_3^\dagger\hat{b}_3^\dagger)|0,0\rangle_{34} \qquad \frac{1}{2}(\hat{b}_3^\dagger\hat{b}_4^\dagger)|0,0\rangle_{34} \qquad -\frac{1}{2}(\hat{b}_3^\dagger\hat{b}_4^\dagger)|0,0\rangle_{34} \qquad \frac{i}{2}(\hat{b}_4^\dagger\hat{b}_4^\dagger)|0,0\rangle_{34}$$

Figure D.6. Illustration showing four possible outputs from an ideal lossless 50:50 beam splitter.

Distinguishability is achieved if the photons have different polarization, different spectral frequency content, different phase, or a time delay between pulses. To quantify how quantum interference is suppressed as the distinguishability of the photons increases, it is convenient to describe interaction with an ideal lossless 50:50 beam splitter using boson creation and annihilation operators.

If there is a single indistinguishable photon at each input port 1 and 2 of the beam splitter, then $|1,1\rangle_{12} = \hat{b}_1^\dagger\hat{b}_2^\dagger |0,0\rangle_{12}$, and, as illustrated in figure D.6, there are four different paths the two photons can take to the output ports 3 and 4. The unitary transform that connects input port states to output port states is given by equation (D.14). The unitary matrix $\hat{U}_B$ has an inverse such that $\hat{U}_B^\dagger = \hat{U}_B^{-1}$ and the creation operator of the superposition output states are such that $\hat{b}_1^\dagger \rightarrow -(i\,\hat{b}_3^\dagger + \hat{b}_4^\dagger)/\sqrt{2}$ and $\hat{b}_2^\dagger \rightarrow -(\hat{b}_3^\dagger + i\hat{b}_4^\dagger)/\sqrt{2}$. The output state is

$$\frac{1}{2}\left(i\,\hat{b}_3^\dagger + \hat{b}_4^\dagger\right)\left(\hat{b}_3^\dagger + i\hat{b}_4^\dagger\right)|0,0\rangle_{34}$$

$$= \frac{1}{2}\left(i\,\hat{b}_3^\dagger\hat{b}_3^\dagger - \hat{b}_3^\dagger\hat{b}_4^\dagger + \hat{b}_3^\dagger\hat{b}_4^\dagger + i\,\hat{b}_4^\dagger\hat{b}_4^\dagger\right)|0,0\rangle_{34} \qquad (D.39)$$

$$= \frac{i}{2}\left(\hat{b}_3^\dagger\hat{b}_3^\dagger + \hat{b}_4^\dagger\hat{b}_4^\dagger\right)|0,0\rangle_{34}$$

in which the terms $-\hat{b}_3^\dagger\hat{b}_4^\dagger + \hat{b}_3^\dagger\hat{b}_4^\dagger$ creating a single photon in each output port exactly cancel. Hence, when a single indistinguishable photon is present at each input port 1 and 2 of the beam splitter, then the only possible output is either two photons at output port 3 or two photons at output port 4. The output state is

$$\frac{i}{2}\left(\hat{b}_3^\dagger\hat{b}_3^\dagger + \hat{b}_4^\dagger\hat{b}_4^\dagger\right)|0,0\rangle_{34} = \frac{i}{2}(|2,0\rangle_{34} + |0,2\rangle_{34}), \qquad (D.40)$$

where use is made of the fact that $\hat{b}_3^\dagger\hat{b}_3^\dagger |0,0\rangle_{34} = \hat{b}_3^\dagger |1,0\rangle_{34} = \sqrt{2} |2,0\rangle_{34}$ and $\hat{b}_4^\dagger\hat{b}_4^\dagger |0,0\rangle_{34} = \hat{b}_4^\dagger |0,1\rangle_{34} = \sqrt{2} |0,2\rangle_{34}$. The probability of detecting two photons at either output port is exactly one-half, and the output port at which the two photons are detected is a fundamentally random quantum mechanical process.

It is possible to continuously tune the distinguishably of identical photons by creating a time delay, τ, in arrival time at the detector. This way, the quantum-to-classical transition can be quantified.

Introducing the time delay at port 2 changes the creation operator to $\hat{b}_2^{\dagger} e^{i\omega\tau}$ and, if the spectral amplitude of each photon pulse is a Gaussian centered at frequency ω_0 with standard deviation σ_0 such that $\phi_0(\omega) = e^{-(\omega-\omega_0)^2/2\sigma_0^2}/(\pi\sigma_0^2)^{\frac{1}{4}}$, it can be shown that the coincidence probability measured by the detectors is [6]

$$\frac{1}{2} - \frac{1}{2}e^{-\sigma_0^2\tau^2/2}. \tag{D.41}$$

Hence, in this case, the 'Mandel dip' in detected photon coincidence counts measured as a function of delay is described using a Gaussian. A typical value of delay that characterizes the extent of the dip in coincidence counts is $\tau = 1$ ps, corresponding to a photon propagating 300 μm in free space.

The wave–particle duality of indistinguishable photons, manifesting as wave-like interference and a measured particle-like 'click' output from a detector, is a purely quantum phenomenon with no classical counterpart.

However, in a somewhat contrived experiment, it *is* possible to configure a classical (or a single photon) analog of the 'Mandel dip' by applying external controls to pseudo randomly switch the quadrature phase difference of classical electromagnetic radiation (or a single photon field) entering the respective input ports of an ideal lossless 50:50 beam splitter such that the output field is multiplexed to appear at either one, but not both, output ports [7]. So, in this sense, any claim of measuring quantum interference and correlation associated with the Mandel dip requires specifying the absence of special external phase modulation of input fields or replacing single-photon detectors with photon-number-resolving detectors.

To see how interference can be used to multiplex two input signals into a single output port, recall that reflection amplitude at a perfect, lossless, symmetric, 50:50 beam splitter is $r_{ph} = -1/\sqrt{2}$ and transmission is $t_{ph} = i/\sqrt{2}$ (equations (D.12) and (D.13)). Flux conservation in the lossless system requires $|r_{ph}|^2 + |t_{ph}|^2 = 1$. If the field amplitude at input port 1 is 1 and at port 2 it is in phase quadrature with a value $-i = t_{ph}/r_{ph}$ then the output field at port 3 is zero and the output field at port 4 is $\sqrt{2}$. The output field can be multiplexed by switching the phase of the port 1 and port 2 input fields. As described in appendix E, a similar approach that exploits interference can be applied to control transient single-photon dynamics in a Fabry–Perot resonator [8].

References

[1] Agnesi A and Degiorio V 2017 *Opt. Laser Tech.* **95** 72
    Degiorio V 1980 *Am. J. Phys.* **48** 81
[2] Ra Y-S, Tichy M C, Lim H-T, Kwon O, Mintert F, Buchleitner A and Kim Y-H 2013 *Proc. Natl Acad. Sci.* **110** 1227
[3] Hong C K, Ou Z Y and Mandel L 1987 *Phys. Rev. Lett.* **59** 2044

[4] Pittman T B, Strekalov D V, Migdall A, Rubin M H, Sergienko A V and Shih Y H 1996 *Phys. Rev. Lett.* **77** 1917

[5] Lalöe F and Mullin W J 2012 *Found. Phys.* **42** 53

[6] Drago C and Brańczyk A M 2024 *Can. J. Phys.* **102** 411

[7] Sadana S, Ghosh D, Joarder K, Lakshmi A N, Sanders B C and Sinha U 2019 *Phys. Rev. A* **100** 013839

[8] Levi A F J, Venuti L C, Albash T and Haas S 2014 *Phys. Rev. A* **90** 022119

IOP Publishing

Essential Semiconductor Laser Device Physics (Second Edition)

A F J Levi

Appendix E

Coherent control of photon dynamics in a Fabry–Perot resonator

E.1 Transmission and reflection of a single photon at a Fabry–Perot resonator

As described in section D.5 of appendix D, controlling the amplitude and phase of coherent light at each of the two input ports of a beam splitter may be used to steer electromagnetic energy to one of the output ports. The relative phase of the field can determine at which output port the light appears, and in this way, the system can act as a multiplexer. Such an approach may be applied to a system consisting of two beam splitters configured as mirrors of a Fabry–Perot resonator and used to control the time evolution of electromagnetic energy density in the optical resonator. The method is an effective way to control the energy density of either classical light or a single photon in an optical resonator. The equations describing the classical and single-photon fields are the same; however, the magnitude squared of the single-photon field, $|\psi(x, t)|^2$, is interpreted as photon energy density [1].

The ability to control photon energy density in an optical resonator is of both fundamental and practical interest. Storage of a photon for a controlled amount of time implies non-Markovian behavior. An example of a practical application is using resonators to delay light in classical communication systems. The control and storage of single photons in a resonator also have potential applications to quantum communication protocols. There is, therefore, motivation to demonstrate control of photon field transient dynamics and, hence, control of transient response in a photon resonator.

Figure E.1 is a schematic of a symmetric one-dimensional Fabry–Perot resonator that consists of two identical lossless dielectric mirrors separated by a cavity of length L_C. Each dielectric mirror has a refractive index n_r and quarter-wave thickness $L_m = \lambda_0/4n_r$, where λ_0 is the resonant photon wavelength in vacuum and the resonant photon frequency is $\omega_0 = 2\pi c/\lambda_0$. Also shown is a photon pulse

doi:10.1088/978-0-7503-6417-1ch12

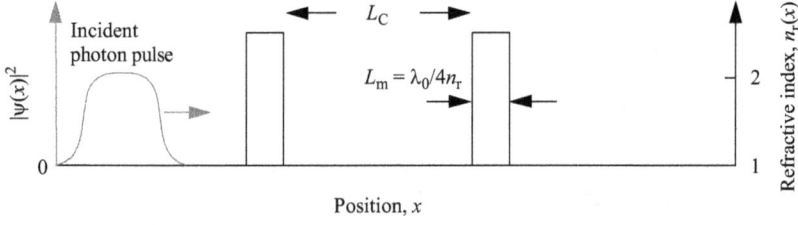

Figure E.1. Illustration of a photon pulse (red curve) with energy density $|\psi(x)|^2$ moving left-to-right and incident on a Fabry–Perot resonator. The ideal lossless dielectric mirrors each have a thickness $L_m = \lambda_0/4n_r$, where n_r is the refractive index, and λ_0 is the resonant wavelength.

(red curve) of energy density $|\psi(x)|^2$ moving left-to-right that is incident on the Fabry–Perot resonator.

It will be convenient to adopt a notation in which, at the resonant photon wavelength, the complex mirror reflection amplitude is r_{ph}, and the transmission amplitude is t_{ph}. Conservation of flux in the lossless system requires $|r_{ph}|^2 + |t_{ph}|^2 = 1$. Optical transmission through each dielectric mirror depends weakly on wavelength according to

$$|t_{ph}|^2 = \frac{1}{1 + \left(\frac{k_1^2 - k_2^2}{2k_1k_2}\right)^2 \sin^2(k_2 L_m)}, \tag{E.1}$$

where the propagation constant in vacuum is $k_1 = 2\pi/\lambda$ and in the mirror it is $k_2 = 2\pi n_r/\lambda$. Transmission $|t_{ph}|^2 = 1/2$ at the resonant wavelength, λ_0, when $L_m = \lambda_0/4n_r$ and $n_r = 1 + \sqrt{2}$.

The transient dynamics of the single-photon optical pulse shown schematically in figure E.1 may be described by time-evolving the photon field $\psi(x, t)$ in which $|\psi(x, t)|^2$ is the photon energy density [1]. A phase-coherent integral of linearly polarized basis states $\phi_\omega(x)$ with amplitudes α_ω gives field

$$\psi(x, t) = \int \frac{d\omega}{2\pi} \alpha_\omega \phi_\omega(x) e^{-i\omega t} \tag{E.2}$$

in which each $\phi_\omega(x)$ is a solution of the one-dimensional Helmholtz equation,

$$\frac{d}{dx}\left(\frac{1}{\mu_r(x)}\frac{d}{dx}\phi_\omega(x)\right) + \omega^2 \varepsilon_r(x)\varepsilon_0\mu_0\phi_\omega(x) = 0. \tag{E.3}$$

It is assumed that relative permeability μ_r and relative permittivity ε_r, with corresponding refractive index $n_r = \sqrt{\mu_r}\sqrt{\varepsilon_r}$, is used to characterize the spatial profile of piecewise-constant lossless dielectric material. At the interface between the jth and $(j+1)$th region of dielectric, the boundary conditions are

$$\phi_{\omega, j}(x_0) = \phi_{\omega, (j+1)}(x_0) \tag{E.4}$$

and

$$\frac{1}{\mu_{r_j}(x)}\frac{d}{dx}\phi_{\omega,j}(x_0) = \frac{1}{\mu_{r_{(j+1)}}(x)}\frac{d}{dx}\phi_{\omega,(j+1)}(x_0).$$ (E.5)

Equation (E.3) describes the system if the coherence time of the photon field is longer than any other characteristic time scale. In practice, the propagation (transfer) matrix method is a convenient way to solve equation (E.3) for the optical resonator illustrated in figure E.1 that is coupled to continuous input and output states [2].

E.2 Transient response

The field of a rectangular photon pulse with center frequency ω_0 and controlled rise and fall time, τ_r, is

$$\psi(x, t) = \int \frac{d\omega}{2\pi}\left(1 + \cos\left(\frac{\pi(\omega - \omega_0)}{\Delta\omega_r}\right)\right)\frac{\sin((\omega - \omega_0)T_0)}{(\omega - \omega_0)T_0}\phi_\omega(x)e^{-i\omega t}.$$ (E.6)

A cosine filters the high-frequency components of the sinc function associated with the rectangular pulse. The rectangular pulse has a duration of $2T_0$ (length $2T_0 c$, where $c = 1/\sqrt{\mu_0\varepsilon_0}$ is the speed of light in vacuum), with a rise and fall time $\tau_r = 2\pi/\Delta\omega_r$.

Typically, the photon energy density of a loaded high-Q optical resonator is assumed to decay as e^{-t/τ_Q} where the time-constant $\tau_Q = 1/\gamma_{ph}$ dominates and is connected via a Fourier transform to a steady-state Lorentzian energy density spectrum,

$$U_{ph}(\omega) = \frac{U_0}{(\omega - \omega_0)^2 + (\gamma_{ph}/2)^2}.$$ (E.7)

However, the transient dynamics of a photon pulse interacting with a Fabry–Perot resonator are more interesting than this. The leading and trailing edges of the optical pulse are spectrally broad, giving rise to non-resonant reflection, and the system evolves in discrete time steps of $\tau_{RT}/2$, where $\tau_{RT} = 2L_C/c$ is the photon round-trip time in the resonator.

As shown in figure E.2(a), the transient dynamics of an optical pulse incident from the left and interacting with a Fabry–Perot resonator can be visualized in a position–time photon energy density plot. The structure imposed in reflected and transmitted energy density occurs in time steps of duration τ_{RT}. The burst of reflected energy at the incident pulse's leading edge and trailing edge is due to non-resonant frequency components contained in the pulse edges.

Figure E.2(b) is a plot of energy density probability $|\psi(x_R, t)|^2$ as a function of time at a position x_R that is far to the right of the resonator. The energy density in the transmitted pulse increases or decreases in a stepwise fashion at the resonant cavity photon round-trip time, τ_{RT}. When centered on the resonance frequency, ω_0, increasing rectangular pulse duration causes the energy density in the Fabry–Perot

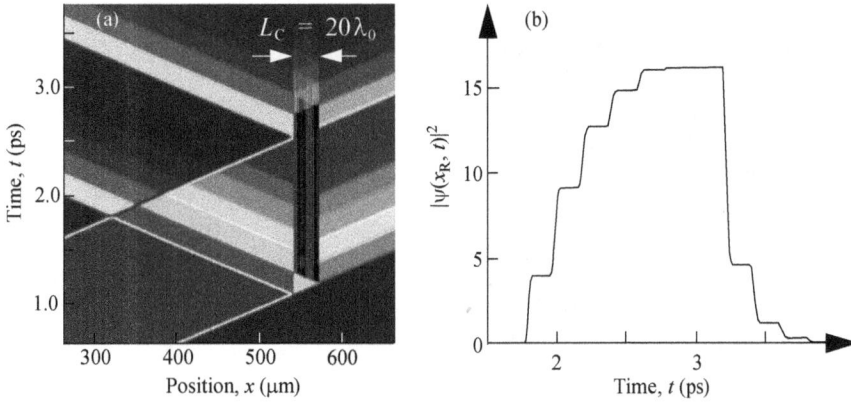

Figure E.2. (a) Rectangular photon pulse incident from the left on a Fabry–Perot resonator shown as a position–time energy density plot. The two arrows indicate the resonator cavity length that has a value $L_C = 20 \times \lambda_0$. (b) Energy density $|\psi(x_R, t)|^2$ (arbitrary scale) detected at position x_R to the right of the resonator. The stepwise change of $|\psi(x_R, t)|^2$ in the resonator occurs at the cavity round-trip time $\tau_{RT} = 40 \times \tau_0 = 200$ fs. The parameters are $\lambda_0 = 1500$ nm, $n_r = 2.5$, $L_m = \lambda_0/4n_r$, $\tau_0 = 5$ fs, $\omega_0 = 2\pi/\tau_0$ rad s^{-1}, $2\Delta\omega_r = \omega_0/4$ rad s^{-1}, and $\omega_0 T_0 = 900$.

to asymptotically approach unity transmission, zero reflection, and maximum energy density in the resonator.

An efficient way to exert temporal control over photon energy density in the Fabry–Perot cavity is through coherent interference effects that exploit the wave nature of the photon. To study how this may be used to control the system response on a time scale as short as the cavity transit time, $\tau_{RT}/2$, it is convenient to consider a photon pulse whose duration is short such that $2T_0 < \tau_{RT}/2$. This helps interpretation after finding the solution to the one-dimensional Helmholtz equation and displaying the resulting photon energy density in a position–time plot. Also helpful is the use of position–time resonant optical ray tracing to estimate reflected and transmitted photon amplitudes at the mirrors of the Fabry–Perot cavity. While ray tracing is an abstraction that approximates the solution to the Helmholtz equation, in practice, it is a very efficient way to find the value of experimentally accessible control parameters that best satisfy a desired objective.

Figure E.3(a) shows the results of calculating the position–time photon energy density of a short pulse moving left-to-right and incident on a Fabry–Perot resonator. A portion of the photon pulse is reflected, and a portion is transmitted through the first dielectric mirror it encounters. The wave character of the photon pulse becomes evident only after reflection from the second mirror it encounters. Self-interference effects occur, and the photon can begin to experience the presence of resonances in the cavity. The photon pulse must interact with both mirrors before it can respond to the presence of the cavity. This is the first step in optical cavity formation. Absent any control, and as shown in the figure, the photon energy injected into the cavity can leak out as forward and backscattered pulses on a time scale set by the cavity transit time $\tau_{RT}/2$. This ring-down in energy density consists of pulses leaving the resonator with a peak value that decays temporally as e^{-t/τ_Q}.

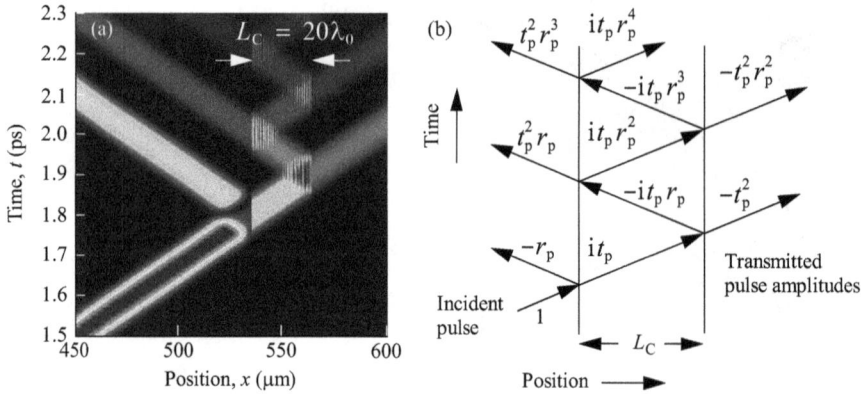

Figure E.3. (a) Ring-down of a short rectangular pulse incident on a Fabry–Perot resonator shown as a position–time energy density plot. (b) Position–time resonant optical ray trace of ring-down showing transmitted and reflected amplitudes. *Note*, for clarity and ease of evaluation, $r_{ph} = -r_p$ and $t_{ph} = it_p$ where r_p and t_p are real. The parameters are $\lambda_0 = 1500$ nm, $L_C = 20 \times \lambda_0$, $n_r = 2.5$, $L_m = \lambda_0/4n_r$, $\tau_0 = 5$ fs, $\omega_0 = 2\pi/\tau_0$ rad s^{-1}, $2\Delta\omega_r = \omega_0/4$ rad s^{-1}, and $\omega_0 T_0 = 60$.

Figure E.3(b) is a position–time resonant optical ray-tracing diagram showing the reflection and transmission coefficients. *Note*, for clarity and ease of evaluation, $r_{ph} = -r_p$ and $t_{ph} = i\,t_p$, where r_p and t_p are real. It is clear from the diagram that the scattered amplitudes at each mirror form a geometric series and that the injection of coherent pulses could be used to control photon energy density in the resonator.

E.3 Coherent control of transient response

Often, the dynamics of a system may be controlled by generating and applying a control field. Closed-loop feedback control requires measurement and so can be quite involved. A simpler approach is to use open-loop control. As illustrated in figure E.4, for a quantum system, this involves defining the physical objective, the model-based creation of control parameters, and using a control-field generator that interacts with the quantum system.

A resonant ray-tracing graph can be used to find near-optimal coherent control parameters for the Fabry–Perot system. The ray-tracing method also has the benefit of avoiding formal optimization methods. Forward or backward propagating pulses of a form similar to equation (E.6), but with different amplitudes and time delays selected by ray tracing, can be applied sequentially to control the resonator's energy density as a function of time. Figure E.5 illustrates how effective this can be. In the figure, a single control pulse is an attenuated, delayed, and phase-shifted version of

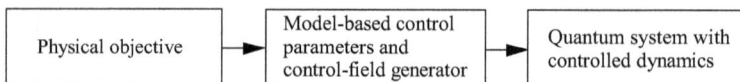

Figure E.4. Illustration of the functional blocks necessary for open-loop control of a quantum system. The control field interacts with the quantum system to control quantum dynamics.

E-5

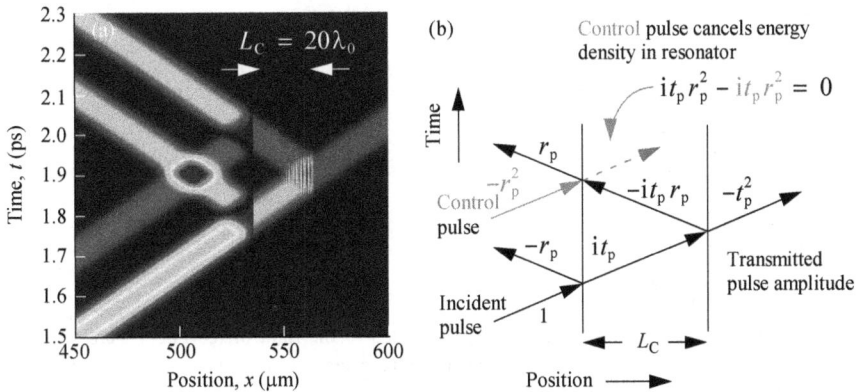

Figure E.5. (a) Incident pulse and control pulse in a position–time photon energy density plot. There is just one transmitted optical pulse after the control pulse removes all electromagnetic energy density in the Fabry–Perot cavity after exactly one round-trip time, τ_{RT}. (b) Position–time resonant optical ray trace showing incident and control amplitudes configured to eliminate ring-down. The parameters are $\lambda_0 = 1500$ nm, $L_C = 20 \times \lambda_0$, $n_r = 2.5$, $L_m = \lambda_0/4n_r$, $\tau_0 = 5$ fs, $\omega_0 = 2\pi/\tau_0$ rad s^{-1}, $2\Delta\omega_r = \omega_0/4$ rad s^{-1}, and $\omega_0 T_0 = 60$.

the incident pulse. In this case, the attenuation is r_p^2, the delay is the cavity round-trip time, τ_{RT}, and the phase shift is π. The effect of this control pulse injected into the cavity at time τ_{RT} after the incident lead pulse is, via interference, to exactly cancel all energy density in the resonator. Figure E.5(a) shows the position–time photon energy density that is the solution to the Helmholtz equation for this system. An incident lead pulse and control pulse move from the left and interact with the Fabry–Perot resonator. The control pulse, with resonant amplitude $-r_p^2$ relative to the incident lead pulse and delayed by a time τ_{RT}, acts to eliminate ring-down after exactly one photon round-trip time in the cavity. Figure E.5(b) illustrates how straightforward it is to find near-optimal sequences of pulses that meet a specific control objective. Subsequently, if required, formal optimization methods can be used to refine the coherent pulse control parameters.

Figure E.3 shows uncontrolled Markovian ring-down of a resonator excited by a single pulse. In this situation, and as shown in figure E.6(a), measurement of the transmitted photon energy density gives a series of pulses at equally spaced time intervals τ_{RT} and with peak value decreasing exponentially as e^{-t/τ_Q}.

Figure E.5 shows a resonator excited by a single pulse and a single coherent control pulse whose effect is to ensure that just one photon pulse is transmitted. The corresponding photon transmission is shown in figure E.6(b). The ring-down evident in figure E.6(a) for the uncontrolled case is eliminated by removing all photon energy density in the cavity after exactly one round-trip time, τ_{RT}, and a single pulse is transmitted.

Coherent control pulses can also propagate in the opposite direction to the incident lead pulse. An example of this is shown in figure E.7, in which the control pulse is used to eliminate ring-down after exactly one transit time, $\tau_{RT}/2$. In this case, there is just one transmitted pulse and one reflected pulse. As is confirmed by

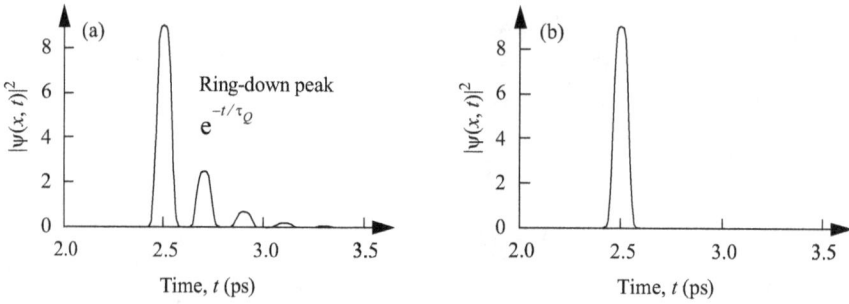

Figure E.6. (a) Transmitted photon energy density evaluated to the right of the Fabry–Perot resonator in figure E.3 as a function of time showing ring-down with no control. (b) The use of a control pulse in (a) eliminates ring-down by removing all photon energy density in the cavity after one round-trip time, τ_{RT}. The parameters are $\lambda_0 = 1500$ nm, $L_C = 20 \times \lambda_0$, $n_r = 2.5$, $L_m = \lambda_0/4n_r$, $\tau_0 = 5$ fs, $\omega_0 = 2\pi/\tau_0$ rad s^{-1}, $2\Delta\omega_r = \omega_0/4$ rad s^{-1}, and $\omega_0 T_0 = 60$. The photon energy density scale is arbitrary.

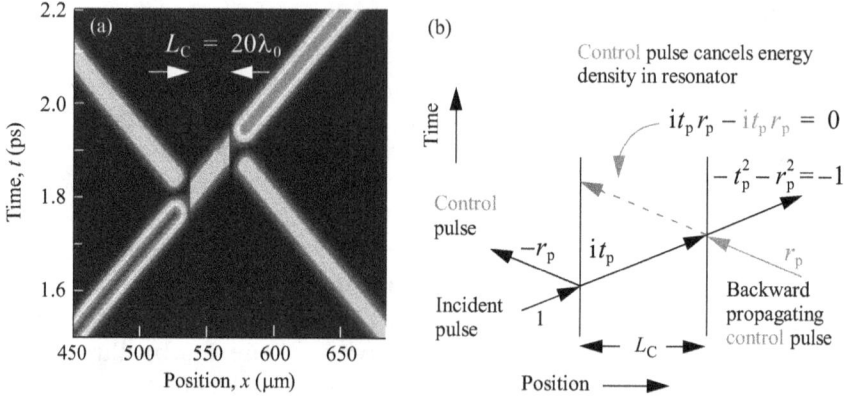

Figure E.7. (a) Position–time photon energy density with an incident lead pulse and a backward propagating control pulse to eliminate ring-down in the Fabry–Perot resonator after one transit time, $\tau_{RT}/2$. (b) Position–time resonant optical ray trace showing incident lead and control amplitudes configured to eliminate ring-down. The parameters are $\lambda_0 = 1500$ nm, $L_C = 20 \times \lambda_0$, $n_r = 2.5$, $L_m = \lambda_0/4n_r$, $\tau_0 = 5$ fs, $\omega_0 = 2\pi/\tau_0$ rad s^{-1}, $2\Delta\omega_r = \omega_0/4$ rad s^{-1}, and $\omega_0 T_0 = 60$. The photon energy density scale is arbitrary.

the resonant ray tracing in figure E.7(b), the control pulse can have *exactly* the same energy density as the reflected pulse so that, viewed from a distance, the control pulse appears delayed in time by $\tau_{RT}/2$ relative to the reflected pulse. If only measuring energy density, it is possible that the reflected pulse appears to have advanced in time by an amount $\tau_{RT}/2$ with respect to the control pulse.

A finite geometric sum can be evaluated by placing an integrating detector at the right-hand output of the Fabry–Perot resonator. A coherent control pulse of amplitude $-r_p^{2N}$ injected at the Nth optical round trip is used to terminate the sum. The resonant ray tracing shown in figure E.8(a) illustrates this for the case $N = 2$. The geometric sum that is measured is

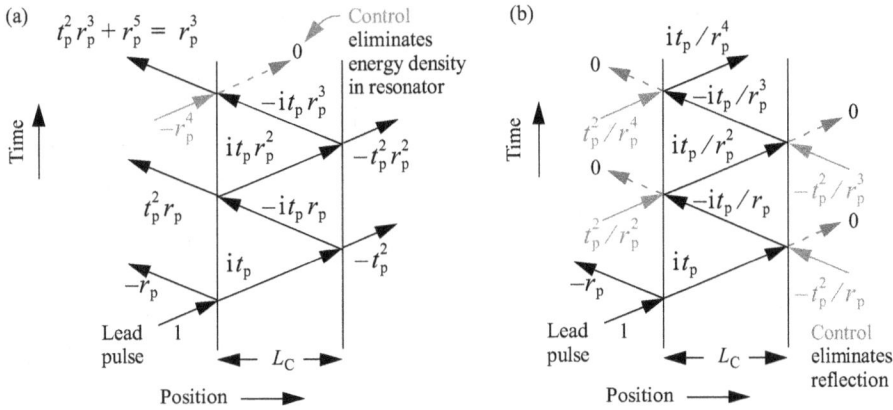

Figure E.8. (a) Position–time resonant optical ray trace showing incident lead and control amplitudes configured to evaluate a finite geometric sum by integrating detection of right-hand output. (b) Position–time resonant optical ray trace showing control amplitudes configured to create a finite divergent geometric series in which electromagnetic energy is controllably confined without any leakage or photon emission from the resonator

$$\left| \sum_{n=0}^{N-1} a\xi^n \right|^2 = \left| a\frac{1-\xi^N}{1-\xi} \right|^2, \tag{E.8}$$

where, on resonance, $\xi = r_p^2$ and $a = t_p^2$. The convergence of the sum in equation (E.8) is guaranteed as $N \to \infty$ because $|r_p| < 1$. More generally, properly configured resonators may be used to evaluate arbitrary finite sums of the form

$$\left| \sum_{n=0}^{N-1} a\xi^n \right|^2, \tag{E.9}$$

where a and ξ are determined by control pulses.

Figure E.8(b) illustrates how a diverging finite geometric sum may be formed using forward- and reverse-propagating control pulses. Coherent optical control pulses are used to *confine* (store) classical electromagnetic energy density for a finite time inside the resonator without any leakage or photon emission from the resonator. During application of the control pulses, *no* electromagnetic energy escapes the resonator, and the classical energy density in the resonator increases (it is amplified) as predicted by equation (E.8) when $|r_p| < 1$. In this way, classical electromagnetic energy can be stored and released from the resonator precisely and deterministically using coherent control pulses [3]. Because the system is linear, it follows that both the input and control signals are not limited to the pulses considered here, and so other control waveforms can be used to achieve the same objective [4].

It can also be shown that a minimum total coherent control energy is required to maintain (store) the photon pulse energy in the resonator to the initially injected value. In this case, and unlike the situation illustrated in figure E.8(b), photon control pulse energy is reflected off the mirrors. For the symmetric resonator illustrated in figure E.8(b), the minimum energy required *per mirror* to ensure the photon energy initially injected into the cavity remains the same and so stored is the energy of the incident pulse multiplied by $|r_p + 1|^2$.

Confinement of a single photon using coherent input and control pulses generated from the same photon requires a modified approach and a different interpretation. With each additional control pulse, photon energy density is increasingly confined to within the resonator. However, because there is only one photon in the system, the classical concept of amplification does not take place. Rather, the single-photon energy density in the resonator increases while the control pulses (whose source is the same photon) are applied. In the presence of noise, there is a limit to the single-photon storage time.

E.4 Optical switching

Communication systems can use laser light sources and glass fiber to transmit data over both short (a few m) and long (many km) distances at data rates that exceed tens and even hundreds of Gb s^{-1}. In these systems, bit periods are often on the scale of 10 ps. For example, a transmission system with a 40 Gb s^{-1} NRZ data channel has a bit period of 25 ps. Typically data are transmitted in packet form with a packet header containing destination information that can be used for routing through a communication network containing fiber-links, packet switches, and destination endpoints. Conventional packet switches convert data from optical to electronic form; data are buffered and switched electronically and then converted back to optical signals for continued transmission through the fiber network. The conversion of optical signals to electronic signals and then back to optical signals adds latency to the communication path and is energy-inefficient compared to maintaining signals in the optical domain. There is, therefore, motivation to create fiber-optic communication networks in which electronic switching is replaced with optical switching. The infrequent reconfiguration of an optical network topology can take place over many bit periods without significantly impacting performance and so this may be achieved relatively easily via all-optical circuit switching. However, all-optical packet switching is more challenging because, absent an efficient all-optical buffer, individual packet headers must be read and acted upon within the bit period, which can be on a 10 ps time scale. In such systems, packet headers should be read, and routing logic operations should be performed on optical signals in real-time. As a starting point for such capability, state machines acting on NRZ optical data would require components that support all-optical Boolean logic functions. In fact, it may be shown that Boolean logic functions such as NAND, NOT, OR, XOR, and XNOR can, in principle, be performed using coherent optical signals and coherent optical control pulses interacting with a beam splitter or a Fabry–Perot resonator. Differential inversion and multiplexing are also possible.

Practical considerations for implementing all-optical packet data routing in systems impose additional requirements. This includes photonic logic circuits that switch fast, are robust against noise, and can be scaled up both in number of devices and logic complexity. In a system using classical coherent light, it is known that the ability to reshape, re-time, and re-amplify optical signals is important for both system function and robustness [5]. Reshaping and re-timing are functions that may be used to remove amplitude and phase noise in a clocked system. Re-amplification of optical signals is needed to drive multiple inputs (fan-out) and to compensate for optical loss. Existing material properties and known device geometries make these '3Rs' difficult to realize in chip-scale optical integrated circuits. The 3Rs are also difficult to implement in an energy-efficient way. Additional helpful functions in such systems are chip-scale all-optical memory and programmable optical delay.

Reshaping and re-timing classical light is possible using a saturable absorber, and re-amplification may be achieved by employing an active optical gain medium. In this situation, the number of photons involved is large enough that the amplification of an electromagnetic field may be considered a smooth and continuous function. This is not something that can be attributed to the single-photon case, and it is for this reason that the single-photon version of 3Rs is more challenging both conceptually and at the device level. For example, amplifying, attenuating, or filtering a single photon state is not possible in the continuum classical sense. A single photon is an indivisible elementary particle, so any amplification might reasonably imply a process in which one or more additional discrete photons are generated. Ideally, for such amplification, each of the discrete number of additional photons would be identical and indistinguishable from the original photon. Alternatively, and from a different perspective illustrated in figure E.8(b), the amplification of a single photon might be interpreted as meaning a continuum increase in *local* photon probability density. In addition to these issues, when one or a few photons in a pure, mixed, or entangled state pass through a beam splitter or filter, quantum noise is introduced, which is often difficult to cancel in practical system implementations.

References

[1] Bialynicki-Birula I 1994 *Acta Phys. Pol.* **86** 97
    Smith B J and Raymer M G 2007 *New J. Phys.* **9** 414
    Raymer M G and Polakos P 2023 *Acta Phys. Pol.* A **143** S28
[2] Levi A F J 2023 *Applied Quantum Mechanics* 3rd edn (Cambridge: Cambridge University Press)
[3] Levi A F J, Venuti L C, Albash T and Haas S 2014 *Phys. Rev.* A **90** 022119
[4] Baranov D G, Krasnok A and Alu A 2017 *Optica* **4** 1457
[5] Horvath T, Radil J, Munster P and Bao N-H 2020 *Appl. Sci.* **10** 5912
    Anagha E G and Jeyachitra R K 2022 *Opt. Eng.* **61** 060902
    Sirleto L and Righini G C 2023 *Micromachines* **14** 614